U0149430

C 语言网络编程实践（影印版）
Hands-On Network
Programming with C

Lewis Van Winkle 著

南京　东南大学出版社

图书在版编目(CIP)数据

C 语言网络编程实践= Hands-On Network
Programming with C:英文/(美)刘易斯·范.温克尔
(Lewis Van Winkle)著. —影印本. —南京:东南大学出版
社,2020.8(2024.8重印)

ISBN 978－7－5641－8956－3

Ⅰ.①C… Ⅱ.①刘… Ⅲ.①C 语言-程序设计-
英文 Ⅳ.①TP312.8

中国版本图书馆 CIP 数据核字(2020)第 109692 号
图字:10－2020－159 号

© 2019 by PACKT Publishing Ltd.

Reprint of the English Edition, jointly published by O'Reilly Media, Inc. and Southeast University
Press, 2020. Authorized reprint of the original English edition, 2020 PACKT Publishing Ltd, the owner
of all rights to publish and sell the same.

All rights reserved including the rights of reproduction in whole or in part in any form.

英文原版由 PACKT Publishing Ltd 出版 2019。

英文影印版由东南大学出版社出版 2020。此影印版的出版和销售得到出版权和销售权的所
有者———PACKT Publishing Ltd 的许可。

版权所有,未得书面许可,本书的任何部分和全部不得以任何形式重制。

C 语言网络编程实践(影印版)

出版发行：东南大学出版社
地　　址：南京四牌楼 2 号　　邮编：210096
出 版 人：江建中
网　　址：http://www.seupress.com
电子邮件：press@seupress.com
印　　刷：广东虎彩云印刷有限公司
开　　本：787 毫米×980 毫米　　16 开本
印　　张：30
字　　数：588 千字
版　　次：2020 年 8 月第 1 版
印　　次：2024 年 8 月第 2 次印刷
书　　号：ISBN 978－7－5641－8956－3
定　　价：158.00 元

本社图书若有印装质量问题,请直接与营销部联系。电话(传真)：025－83791830

For Doogie

– Lewis Van Winkle

`mapt.io`

Mapt is an online digital library that gives you full access to over 5,000 books and videos, as well as industry leading tools to help you plan your personal development and advance your career. For more information, please visit our website.

Why subscribe?

- Spend less time learning and more time coding with practical eBooks and Videos from over 4,000 industry professionals

- Improve your learning with Skill Plans built especially for you

- Get a free eBook or video every month

- Mapt is fully searchable

- Copy and paste, print, and bookmark content

Packt.com

Did you know that Packt offers eBook versions of every book published, with PDF and ePub files available? You can upgrade to the eBook version at `www.packt.com` and as a print book customer, you are entitled to a discount on the eBook copy. Get in touch with us at `customercare@packtpub.com` for more details.

At `www.packt.com`, you can also read a collection of free technical articles, sign up for a range of free newsletters, and receive exclusive discounts and offers on Packt books and eBooks.

Contributors

About the author

Lewis Van Winkle is a software programming consultant, entrepreneur, and founder of a successful IoT company. He has over 20 years of programming experience after publishing his first successful software product at the age of 12. He has over 15 years of programming experience with the C programming language on a variety of operating systems and platforms. He is active in the open source community and has published several popular open source programs and libraries—many of them in C. Today, Lewis spends much of his time consulting, where he loves taking on difficult projects that other programmers have given up on. He specializes in network systems, financial systems, machine learning, and interoperation between different programming languages.

I would like to thank the publisher, Packt. This book wouldn't exist without their encouragement and backing. I would also like to extend a special thank you to my reviewer, Daniele Lacamera, for the careful work he carried out. This book improved significantly as a result of his valuable feedback. I also want to acknowledge the patience and support that my friends and family have shown over the last year while I've been away writing.

About the reviewer

Daniele Lacamera is a software technologist and researcher with vast experience in software design and development on embedded systems for different industries. He is currently working as freelance software developer and trainer. He is a worldwide expert in TCP/IP and transport protocol design and optimization, with more than 20 academic publications on the topic. He supports free software by contributing to several projects, including the Linux kernel, and is involved within a number of communities and organizations that promote the use of free and open source software in the IoT.

Packt is searching for authors like you

If you're interested in becoming an author for Packt, please visit `authors.packtpub.com` and apply today. We have worked with thousands of developers and tech professionals, just like you, to help them share their insight with the global tech community. You can make a general application, apply for a specific hot topic that we are recruiting an author for, or submit your own idea.

Table of Contents

Preface

Packt first contacted me about writing this book nearly a year ago. It's been a long journey, harder than I anticipated at times, and I've learned a lot. The book you hold now is the culmination of many long days, and I'm proud to finally present it.

I think C is a beautiful programming language. No other language in everyday use gets you as close to the machine as C does. I've used C to program 8-bit microcontrollers with only 16 bytes of RAM, just the same as I've used it to program modern desktops with multi-core, multi-GHz processors. It's truly remarkable that C works efficiently in both contexts.

Network programming is a fun topic, but it's also a very deep one; a lot is going on at many levels. Some programming languages hide these abstractions. In the Python programming language, for example, you can download an entire web page using only one line of code. This isn't the case in C! In C, if you want to download a web page, you have to know how everything works. You need to know sockets, you need to know **Transfer Control Protocol (TCP)**, and you need to know HTTP. In C network programming, nothing is hidden.

C is a great language to learn network programming in. This is not only because we get to see all the details, but also because the popular operating systems all use kernels written in C. No other language gives you the same first-class access as C does. In C, everything is under your control – you can lay out your data structures exactly how you want, manage memory precisely as you please, and even shoot yourself in the foot just the way you want.

When I first began writing this book, I surveyed other resources related to learning network programming with C. I found much misinformation – not only on the web, but even in print. There is a lot of C networking code that is done wrong. Internet tutorials about C sockets often use deprecated functions and ignore memory safety completely. When it comes to network programming, you can't take the *it works so it's good enough programming-by-coincidence* approach. You have to use reasoning.

In this book, I take care to approach network programming in a modern and safe way. The example programs are carefully designed to work with both IPv4 and IPv6, and they are all written in a portable, operating system-independent way, whenever possible. Wherever there is an opportunity for memory errors, I try to take notice and point out these concerns. Security is too often left as an afterthought. I believe security is important, and it should be planned in the system from the beginning. Therefore, in addition to teaching network basics, this book spends a lot of time working with secure protocols, such as TLS.

I hope you enjoy reading this book as much as I enjoyed writing it.

Who this book is for

This book is for the C or C++ programmer who wants to add networking features to their software. It is also designed for the student or professional who simply wants to learn about network programming and common network protocols.

It is assumed that the reader already has some familiarity with the C programming language. This includes a basic proficiency with pointers, basic data structures, and manual memory management.

What this book covers

Chapter 1, *Introducing Networks and Protocols*, introduces the important concepts related to networking. This chapter includes example programs to determine your IP address pragmatically.

Chapter 2, *Getting to Grips with Socket APIs*, introduces socket programming APIs and has you build your first networked program—a tiny web server.

Chapter 3, *An In-Depth Overview of TCP Connections*, focuses on programming TCP sockets. In this chapter, example programs are developed for both the client and server sides.

Chapter 4, *Establishing UDP Connections*, covers programming with **User Datagram Protocol** (**UDP**) sockets.

Chapter 5, *Hostname Resolution and DNS*, explains how hostnames are translated into IP addresses. In this chapter, we build an example program to perform manual DNS query lookups using UDP.

Chapter 6, *Building a Simple Web Client*, introduces HTTP—the protocol that powers websites. We dive right in and build an HTTP client in C.

Chapter 7, *Building a Simple Web Server*, describes how to construct a fully functional web server in C. This program is able to serve a static website to any modern web browser.

Chapter 8, *Making Your Program Send Email*, describes **Simple Mail Transfer Protocol** (**SMTP**)—the protocol that is powering email. In this chapter, we develop a program that can send email over the internet.

Chapter 9, *Loading Secure Web Pages with HTTPS and OpenSSL*, explores TLS—the protocol that secures web pages. In this chapter, we develop an HTTPS client that is capable of downloading web pages securely.

Chapter 10, *Implementing a Secure Web Server*, continues on the security theme and explores the construction of a secure HTTPS web server.

Chapter 11, *Establishing SSH Connections with libssh*, continues with the secure protocol theme. The use of **Secure Shell** (**SSH**) is covered to connect to a remote server, execute commands, and download files securely.

Chapter 12, *Network Monitoring and Security*, discusses the tools and techniques used to test network functionality, troubleshoot problems, and eavesdrop on insecure communication protocols.

Chapter 13, *Socket Programming Tips and Pitfalls*, goes into detail about TCP and addresses many important edge cases that appear in socket programming. The techniques covered are invaluable for creating robust network programs.

Chapter 14, *Web Programming for the Internet of Things*, gives an overview of the design and programming for **Internet of Things** (**IoT**) applications.

Appendix A, *Answers to Questions*, provides answers to the comprehension questions given at the end of each chapter.

Appendix B, *Setting Up Your C Compiler on Windows*, gives instructions for setting up a development environment on Windows that is needed for compiling all of the example programs in this book.

Appendix C, *Setting Up Your C Compiler on Linux*, provides the setup instructions for preparing your Linux computer to be capable of compiling all of the example programs in this book.

Appendix D, *Setting Up Your C Compiler on macOS*, gives step-by-step instructions for configuring your macOS system to be capable of compiling all of the example programs in this book.

Appendix E, *Example Programs*, lists each example program, by chapter, included in this book's code repository.

To get the most out of this book

The reader is expected to be proficient in the C programming language. This includes a familiarity with memory management, the use of pointers, and basic data structures.

A Windows, Linux, or macOS development machine is recommended; you can refer to the appendices for setup instructions.

This book takes a hands-on approach to learning and includes 44 example programs. Working through these examples as you read the book will help enforce the concepts.

The code for this book is released under the MIT open source license. The reader is encouraged to use, modify, improve, and even publish their changes to these example programs.

Download the example code files

You can download the example code files for this book from your account at `www.packt.com`. If you purchased this book elsewhere, you can visit `www.packt.com/support` and register to have the files emailed directly to you.

The code bundle for the book is also publicly hosted on GitHub at `https://github.com/codeplea/hands-on-network-programming-with-c`. In case there's an update to the code, it will be updated on that GitHub repository. Each chapter that introduces example programs begins with the commands needed to download the book's code.

Download the color images

We also provide a PDF file that has color images of the screenshots/diagrams used in this book. You can download it here: `http://www.packtpub.com/sites/default/files/downloads/9781789349863_ColorImages.pdf`.

Conventions used

There are a number of text conventions used throughout this book.

`CodeInText`: Indicates code words in text, variable names, function names, directory names, filenames, file extensions, pathnames, URLs, and user input. Here is an example: "Use the `select()` function to wait for network data."

A block of code is set as follows:

```
/* example program */

#include <stdio.h>
int main() {
    printf("Hello World!\n");
    return 0;
}
```

Any command-line input or output is written as follows:

```
gcc hello.c -o hello
./hello
```

Bold: Indicates a new term, an important word, or words that you see on screen. For example, words in menus or dialog boxes appear in the text like this. Here is an example: "Select **System info** from the **Administration** panel."

Get in touch

Feedback from our readers is always welcome.

General feedback: If you have questions about any aspect of this book, mention the book title in the subject of your message and email us at customercare@packtpub.com.

Errata: Although we have taken every care to ensure the accuracy of our content, mistakes do happen. If you have found a mistake in this book, we would be grateful if you would report this to us. Please visit www.packt.com/submit-errata, selecting your book, clicking on the Errata Submission Form link, and entering the details.

Piracy: If you come across any illegal copies of our works in any form on the internet, we would be grateful if you would provide us with the location address or website name. Please contact us at copyright@packt.com with a link to the material.

If you are interested in becoming an author: If there is a topic that you have expertise in, and you are interested in either writing or contributing to a book, please visit authors.packtpub.com.

Reviews

Please leave a review. Once you have read and used this book, why not leave a review on the site that you purchased it from? Potential readers can then see and use your unbiased opinion to make purchase decisions, we at Packt can understand what you think about our products, and our authors can see your feedback on their book. Thank you!

For more information about Packt, please visit `packt.com`.

Section 1 - Getting Started with Network Programming

This section will get the reader up and running with the basics of networking, the relevant network protocols, and basic socket programming.

The following chapters are in this section:

Introducing Networks and Protocols

1

In this chapter, we will review the fundamentals of computer networking. We'll look at abstract models that attempt to explain the main concerns of networking, and we'll explain the operation of the primary network protocol, the Internet Protocol. We'll look at address families and end with writing programs to list your computer's local IP addresses.

The following topics are covered in this chapter:

- Network programming and C
- OSI layer model
- TCP/IP reference model
- The Internet Protocol
- IPv4 addresses and IPv6 addresses
- Domain names
- Internet protocol routing
- Network address translation
- The client-server paradigm
- Listing your IP addresses programmatically from C

Technical requirements

Most of this chapter focuses on theory and concepts. However, we do introduce some sample programs near the end. To compile these programs, you will need a good C compiler. We recommend MinGW on Windows and GCC on Linux and macOS. See `Appendix B`, *Setting Up Your C Compiler On Windows*, `Appendix C`, *Setting Up Your C Compiler On Linux*, and `Appendix D`, *Setting Up Your C Compiler On macOS*, for compiler setup.

The code for this book can be found at: `https://github.com/codeplea/Hands-On-Network-Programming-with-C`.

From the command line, you can download the code for this chapter with the following command:

```
git clone https://github.com/codeplea/Hands-On-Network-Programming-with-C
cd Hands-On-Network-Programming-with-C/chap01
```

On Windows, using MinGW, you can use the following command to compile and run code:

```
gcc win_list.c -o win_list.exe -liphlpapi -lws2_32
win_list
```

On Linux and macOS, you can use the following command:

```
gcc unix_list.c -o unix_list
./unix_list
```

The internet and C

Today, the internet needs no introduction. Certainly, millions of desktops, laptops, routers, and servers are connected to the internet and have been for decades. However, billions of additional devices are now connected as well—mobile phones, tablets, gaming systems, vehicles, refrigerators, television sets, industrial machinery, surveillance systems, doorbells, and even light bulbs. The new **Internet of Things** (**IoT**) trend has people rushing to connect even more unlikely devices every day.

Over 20 billion devices are estimated to be connected to the internet now. These devices use a wide variety of hardware. They connect over an Ethernet connection, Wi-Fi, cellular, a phone line, fiber optics, and other media, but they likely have one thing in common; they likely use **C**.

The use of the C programming language is ubiquitous. Almost every network stack is programmed in C. This is true for Windows, Linux, and macOS. If your mobile phone uses Android or iOS, then even though the apps for these were programmed in a different language (Java and Objective C), the kernel and networking code was written in C. It is very likely that the network routers that your internet data goes through are programmed in C. Even if the user interface and higher-level functions of your modem or router are programmed in another language, the networking drivers are still probably implemented in C.

Networking encompasses concerns at many different abstraction levels. The concerns your web browser has with formatting a web page are much different than the concerns your router has with forwarding network packets. For this reason, it is useful to have a theoretical model that helps us to understand communications at these different levels of abstraction. Let's look at these models now.

OSI layer model

It's clear that if all of the disparate devices composing the internet are going to communicate seamlessly, there must be agreed-upon standards that define their communications. These standards are called **protocols**. Protocols define everything from the voltage levels on an Ethernet cable to how a JPEG image is compressed on a web page. It's clear that, when we talk about the voltage on an Ethernet cable, we are at a much different level of abstraction compared to talking about the JPEG image format. If you're programming a website, you don't want to think about Ethernet cables or Wi-Fi frequencies. Likewise, if you're programming an internet router, you don't want to have to worry about how JPEG images are compressed. For this reason, we break the problem down into many smaller pieces.

One common method of breaking down the problem is to place levels of concern into layers. Each layer then provides services for the layer on top of it, and each upper layer can rely on the layers underneath it without concern for how they work.

The most popular layer system for networking is called the **Open Systems Interconnection** model (**OSI** model). It was standardized in 1977 and is published as ISO 7498. It has seven layers:

7	Application
6	Presentation
5	Session
4	Transport
3	Network
2	Data Link
1	Physical

Let's understand these layers one by one:

- **Physical** (1): This is the level of physical communication in the real world. At this level, we have specifications for things such as the voltage levels on an Ethernet cable, what each pin on a connector is for, the radio frequency of Wi-Fi, and the light flashes over an optic fiber.
- **Data Link** (2): This level builds on the physical layer. It deals with protocols for directly communicating between two nodes. It defines how a direct message between nodes starts and ends (framing), error detection and correction, and flow control.
- **Network layer** (3): The network layer provides the methods to transmit data sequences (called packets) between nodes in different networks. It provides methods to route packets from one node to another (without a direct physical connection) by transferring through many intermediate nodes. This is the layer that the Internet Protocol is defined on, which we will go into in some depth later.
- **Transport layer** (4): At this layer, we have methods to reliably deliver variable length data between hosts. These methods deal with splitting up data, recombining it, ensuring data arrives in order, and so on. The **Transmission Control Protocol** (**TCP**) and **User Datagram Protocol** (**UDP**) are commonly said to exist on this layer.
- **Session layer** (5): This layer builds on the transport layer by adding methods to establish, checkpoint, suspend, resume, and terminate dialogs.
- **Presentation layer** (6): This is the lowest layer at which data structure and presentation for an application are defined. Concerns such as data encoding, serialization, and encryption are handled here.
- **Application layer** (7): The applications that the user interfaces with (for example, web browsers and email clients) exist here. These applications make use of the services provided by the six lower layers.

In the OSI model, an application, such as a web browser, exists in the application layer (layer 7). A protocol from this layer, such as HTTP used to transmit web pages, doesn't have to concern itself with how the data is being transmitted. It can rely on services provided by the layer underneath it to effectively transmit data. This is illustrated in the following diagram:

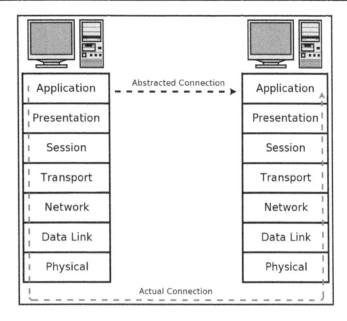

It should be noted that chunks of data are often referred to by different names depending on the OSI layer they're on. A data unit on layer 2 is called a **frame**, since layer 2 is responsible for framing messages. A data unit on layer 3 is referred to as a **packet**, while a data unit on layer 4 is a **segment** if it is part of a TCP connection or a **datagram** if it is a UDP message.

In this book, we often use the term packet as a generic term to refer to a data unit on any layer. However, segment will only be used in the context of a TCP connection, and datagram will only refer to UDP datagrams.

As we will see in the next section, the OSI model doesn't fit precisely with the common protocols in use today. However, it is still a handy model to explain networking concerns, and it is still in widespread use for that purpose today.

TCP/IP layer model

The **TCP/IP protocol suite** is the most common network communication model in use today. The TCP/IP reference model differs a bit from the OSI model, as it has only four layers instead of seven.

The following diagram illustrates how the four layers of the TCP/IP model line up to the seven layers of the OSI model:

	OSI Model	TCP/IP Model	
7	Application	Process/ Application	4
6	Presentation		
5	Session		
4	Transport	Host-to-Host	3
3	Network	Internet	2
2	Data Link	Network Access	1
1	Physical		

Notably, the TCP/IP model doesn't match up exactly with the layers in the OSI model. That's OK. In both models, the same functions are performed; they are just divided differently.

The TCP/IP reference model was developed after the TCP/IP protocol was already in common use. It differs from the OSI model by subscribing a less rigid, although still hierarchical, model. For this reason, the OSI model is sometimes better for understanding and reasoning about networking concerns, but the TCP/IP model reflects a more realistic view of how networking is commonly implemented today.

The four layers of the TCP/IP model are as follows:

- **Network Access layer** (1): On this layer, physical connections and data framing happen. Sending an Ethernet or Wi-Fi packet are examples of layer 1 concerns.
- **Internet layer** (2): This layer deals with the concerns of addressing packets and routing them over multiple interconnection networks. It's at this layer that an IP address is defined.
- **Host-to-Host layer** (3): The host-to-host layer provides two protocols, TCP and UDP, which we will discuss in the next few chapters. These protocols address concerns such as data order, data segmentation, network congestion, and error correction.
- **Process/Application layer** (4): The process/application layer is where protocols such as HTTP, SMTP, and FTP are implemented. Most of the programs that feature in this book could be considered to take place on this layer while consuming functionality provided by our operating system's implementation of the lower layers.

Regardless of your chosen abstraction model, real-world protocols do work at many levels. Lower levels are responsible for handling data for the higher levels. These lower-level data structures must, therefore, encapsulate data from the higher levels. Let's look at encapsulating data now.

Data encapsulation

The advantage of these abstractions is that, when programming an application, we only need to consider the highest-level protocol. For example, a web browser needs only to implement the protocols dealing specifically with websites—HTTP, HTML, CSS, and so on. It does not need to bother with implementing TCP/IP, and it certainly doesn't have to understand how an Ethernet or Wi-Fi packet is encoded. It can rely on ready-made implementations of the lower layers for these tasks. These implementations are provided by the operating system (for example, Windows, Linux, and macOS).

When communicating over a network, data must be processed down through the layers at the sender and up again through the layers at the receiver. For example, if we have a web server, **Host A**, which is transmitting a web page to the receiver, **Host B**, it may look like this:

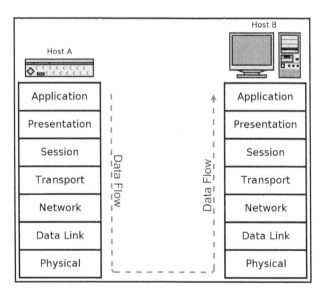

The web page contains a few paragraphs of text, but the web server doesn't only send the text by itself. For the text to be rendered correctly, it must be encoded in an **HTML** structure:

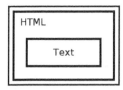

In some cases, the text is already preformatted into **HTML** and saved that way but, in this example, we are considering a web application that dynamically generates the **HTML**, which is the most common paradigm for dynamic web pages. As the text cannot be transmitted directly, neither can the **HTML**. It instead must be transmitted as part of an **HTTP** response. The web server does this by applying the appropriate **HTTP** response header to the **HTML**:

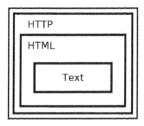

The **HTTP** is transmitted as part of a **TCP** session. This isn't done explicitly by the web server, but is taken care of by the operating system's TCP/IP stack:

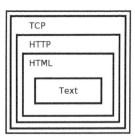

The **TCP** packet is routed by an **IP** packet:

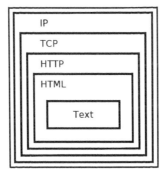

This is transmitted over the wire in an **Ethernet** packet (or another protocol):

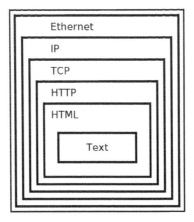

Luckily for us, the lower-level concerns are handled automatically when we use the socket APIs for network programming. It is still useful to know what happens behind the scenes. Without this knowledge, dealing with failures or optimizing for performance is difficult if not impossible.

With some of the theory out of the way, let's dive into the actual protocols powering modern networking.

Internet Protocol

Twenty years ago, there were many competing networking protocols. Today, one protocol is overwhelmingly common—the Internet Protocol. It comes in two versions—IPv4 and IPv6. IPv4 is completely ubiquitous and deployed everywhere. If you're deploying network code today, you must support IPv4 or risk that a significant portion of your users won't be able to connect.

IPv4 uses 32-bit addresses, which limits it to addressing no more than 2^{32} or 4,294,967,296 systems. However, these 4.3 billion addresses were not initially assigned efficiently, and now many **Internet Service Providers (ISPs)** are forced to ration IPv4 addresses.

IPv6 was designed to replace IPv4 and has been standardized by the **Internet Engineering Task Force (IETF)** since 1998. It uses a 128-bit address, which allows it to address a theoretical 2^{128} = 340,282,366,920,938,463,463,374,607,431,768,211,456, or about a 3.4 x 10^{38} addresses.

Today, every major desktop and smartphone operating system supports both IPv4 and IPv6 in what is called a **dual-stack configuration**. However, many applications, servers, and networks are still only configured to use IPv4. From a practical standpoint, this means that you need to support IPv4 in order to access much of the internet. However, you should also support IPv6 to be future-proof and to help the world to transition away from IPv4.

What is an address?

All Internet Protocol traffic routes to an address. This is similar to how phone calls must be dialed to phone numbers. IPv4 addresses are 32 bits long. They are commonly divided into four 8-bit sections. Each section is displayed as a decimal number between 0 and 255 inclusive and is delineated by a period.

Here are some examples of IPv4 addresses:

- 0.0.0.0
- 127.0.0.1
- 10.0.0.0
- 172.16.0.5
- 192.168.0.1
- 192.168.50.1
- 255.255.255.255

A special address, called the **loopback** address, is reserved at `127.0.0.1`. This address essentially means *establish a connection to myself*. Operating systems short-circuit this address so that packets to it never enter the network but instead stay local on the originating system.

IPv4 reserves some address ranges for private use. If you're using IPv4 through a router/NAT, then you are likely using an IP address in one of these ranges. These reserved private ranges are as follows:

- `10.0.0.0` to `10.255.255.255`
- `172.16.0.0` to `172.31.255.255`
- `192.168.0.0` to `192.168.255.255`

The concept of IP address ranges is a useful one that comes up many times in networking. It's probably not surprising then that there is a shorthand notation for writing them. Using **Classless Inter-Domain Routing** (**CIDR**) notation, we can write the three previous address ranges as follows:

- `10.0.0.0/8`
- `172.16.0.0/12`
- `192.168.0.0/16`

CIDR notation works by specifying the number of bits that are fixed. For example, `10.0.0.0/8` specifies that the first eight bits of the `10.0.0.0` address are fixed, the first eight bits being just the first `10.` part; the remaining `0.0.0` part of the address can be anything and still be on the `10.0.0.0/8` block. Therefore, `10.0.0.0/8` encompasses `10.0.0.0` through `10.255.255.255`.

IPv6 addresses are 128 bits long. They are written as eight groups of four hexadecimal characters delineated by colons. A hexadecimal character can be from 0-9 or from a-f. Here are some examples of IPv6 addresses:

- `0000:0000:0000:0000:0000:0000:0000:0001`
- `2001:0db8:0000:0000:0000:ff00:0042:8329`
- `fe80:0000:0000:0000:75f4:ac69:5fa7:67f9`
- `ffff:ffff:ffff:ffff:ffff:ffff:ffff:ffff`

Note that the standard is to use lowercase letters in IPv6 addresses. This is in contrast to many other uses of hexadecimal in computers.

There are a couple of rules for shortening IPv6 addresses to make them easier. Rule 1 allows for the leading zeros in each section to be omitted (for example, `0db8` = `db8`). Rule 2 allows for consecutive sections of zeros to be replaced with a double colon (`::`). Rule 2 may only be used once in each address; otherwise, the address would be ambiguous.

Applying both rules, the preceding addresses can be shortened as follows:

- `::1`
- `2001:db8::ff00:42:8329`
- `fe80::75f4:ac69:5fa7:67f9`
- `ffff:ffff:ffff:ffff:ffff:ffff:ffff:ffff`

Like IPv4, IPv6 also has a loopback address. It is `::1`.

Dual-stack implementations also recognize a special class of IPv6 address that map directly to an IPv4 address. These reserved addresses start with 80 zero bits, and then by 16 one bits, followed by the 32-bit IPv4 address. Using CIDR notation, this block of address is `::ffff:0:0/96`.

These mapped addresses are commonly written with the first 96 bits in IPv6 format followed by the remaining 32 bits in IPv4 format. Here are some examples:

IPv6 Address	Mapped IPv4 Address
`::ffff:10.0.0.0`	`10.0.0.0`
`::ffff:172.16.0.5`	`172.16.0.5`
`::ffff:192.168.0.1`	`192.168.0.1`
`::ffff:192.168.50.1`	`192.168.50.1`

You may also run into IPv6 **site-local addresses**. These site-local addresses are in the `fec0::/10` range and are for use on private local networks. Site-local addresses have now been deprecated and should not be used for new networks, but many existing implementations still use them.

Another address type that you should be familiar with are **link-local addresses**. Link-local addresses are usable only on the local link. Routers never forward packets from these addresses. They are useful for a system to accesses auto-configuration functions before having an assigned IP address. Link-local addresses are in the IPv4 `169.254.0.0/16` address block or the IPv6 `fe80::/10` address block.

It should be noted the IPv6 introduces many additional features over IPv4 besides just a greatly expanded address range. IPv6 addresses have new attributes, such as scope and lifetime, and it is normal for IPv6 network interfaces to have multiple IPv6 addresses. IPv6 addresses are used and managed differently than IPv4 addresses.

Regardless of these differences, in this book, we strive to write code that works well for both IPv4 and IPv6.

If you think that IPv4 addresses are difficult to memorize, and IPv6 addresses impossible, then you are not alone. Luckily, we have a system to assign names to specific addresses.

Domain names

The Internet Protocol can only route packets to an IP address, not a name. So, if you try to connect to a website, such as `example.com`, your system must first resolve that domain name, `example.com`, into an IP address for the server that hosts that website.

This is done by connecting to a **Domain Name System** (**DNS**) server. You connect to a domain name server by knowing in advance its IP address. The IP address for a domain name server is usually assigned by your ISP.

Many other domain name servers are made publicly available by different organizations. Here are a few free and public DNS servers:

DNS Provider	IPv4 Addresses	IPv6 Addresses
Cloudflare 1.1.1.1	`1.1.1.1`	`2606:4700:4700::1111`
	`1.0.0.1`	`2606:4700:4700::1001`
FreeDNS	`37.235.1.174`	
	`37.235.1.177`	
Google Public DNS	`8.8.8.8`	`2001:4860:4860::8888`
	`8.8.4.4`	`2001:4860:4860::8844`
OpenDNS	`208.67.222.222`	`2620:0:ccc::2`
	`208.67.220.220`	`2620:0:ccd::2`

To resolve a hostname, your computer sends a UDP message to your domain name server and asks it for an AAAA-type record for the domain you're trying to resolve. If this record exists, an IPv6 address is returned. You can then connect to a server at that address to load the website. If no AAAA record exists, then your computer queries the server again, but asks for an A record. If this record exists, you will receive an IPv4 address for the server. In many cases, a site will publish an A record and an AAAA record that route to the same server.

It is also possible, and common, for multiple records of the same type to exist, each pointing to a different address. This is useful for redundancy in the case where multiple servers can provide the same service.

We will see a lot more about DNS queries in Chapter 5, *Hostname Resolution and DNS*.

Now that we have a basic understanding of IP addresses and names, let's look into detail of how IP packets are routed over the internet.

Internet routing

If all networks contained only a maximum of only two devices, then there would be no need for routing. Computer A would just send its data directly over the wire, and computer B would receive it as the only possibility:

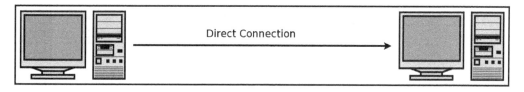

The internet today has an estimated 20 billion devices connected. When you make a connection over the internet, your data first transmits to your local router. From there, it is transmitted to another router, which is connected to another router, and so on. Eventually, your data reaches a router that is connected to the receiving device, at which point, the data has reached its destination:

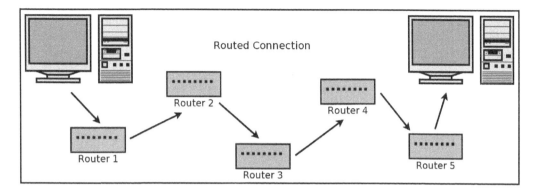

Imagine that each router in the preceding diagram is connected to tens, hundreds, or even thousands of other routers and systems. It's an amazing feat that IP can discover the correct path and deliver traffic seamlessly.

Windows includes a utility, `tracert`, which lists the routers between your system and the destination system.

Here is an example of using the `tracert` command on Windows 10 to trace the route to `example.com`:

```
Windows PowerShell                                            —    □    ×
PS C:\> tracert example.com

Tracing route to example.com [93.184.216.34]
over a maximum of 30 hops:

  1    <1 ms    <1 ms    <1 ms   192.168.50.1
  2     *        *        *      Request timed out.
  3     *        *        *      Request timed out.
  4     2 ms     2 ms     1 ms   my.jetpack [192.168.1.1]
  5   119 ms    47 ms    41 ms   172.26.96.169
  6    66 ms    39 ms    38 ms   107.79.227.124
  7     *        *        *      Request timed out.
  8    58 ms    79 ms    70 ms   12.83.186.145
  9    61 ms    40 ms    41 ms   cgcil403igs.ip.att.net [12.122.133.33]
 10    78 ms    38 ms    39 ms   dcr1-so-4-0-0.atlanta.savvis.net [192.205.32.118]
 11   116 ms   198 ms    47 ms   192.229.225.133
 12    76 ms    40 ms    37 ms   93.184.216.34

Trace complete.
PS C:\>
```

As you can see from the example, there are 11 hops between our system and the destination system (example.com, 93.184.216.34). The IP addresses are listed for many of these intermediate routers, but a few are missing with the Request timed out message. This usually means that the system in question doesn't support the part of the **Internet Control Message Protocol** (**ICMP**) protocol needed. It's not unusual to see a few such systems when running tracert.

In Unix-based systems, the utility to trace routes is called traceroute. You would use it like traceroute example.com, for example, but the information obtained is essentially the same.

More information on tracert and traceroute can be found in Chapter 12, *Network Monitoring and Security*.

Sometimes, when IP packets are transferred between networks, their addresses must be translated. This is especially common when using IPv4. Let's look at the mechanism for this next.

Local networks and address translation

It's common for households and organizations to have small **Local Area Networks** (**LANs**). As mentioned previously, there are IPv4 addresses ranges reserved for use in these small local networks.

These reserved private ranges are as follows:

- 10.0.0.0 to 10.255.255.255
- 172.16.0.0 to 172.31.255.255
- 192.168.0.0 to 192.168.255.255

When a packet originates from a device on an IPv4 local network, it must undergo **Network Address Translation** (**NAT**) before being routed on the internet. A router that implements NAT remembers which local address a connection is established from.

The devices on the same LAN can directly address one another by their local address. However, any traffic communicated to the internet must undergo address translation by the router. The router does this by modifying the source IP address from the original private LAN IP address to its public internet IP address:

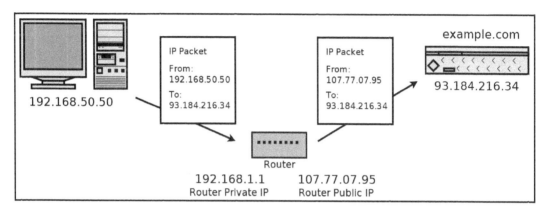

Likewise, when the router receives the return communication, it must modify the destination address from its public IP to the private IP of the original sender. It knows the private IP address because it was stored in memory after the first outgoing packet:

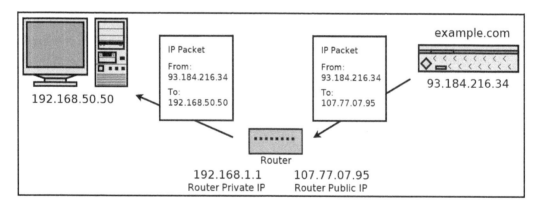

Network address translation can be more complicated than it first appears. In addition to modifying the source IP address in the packet, it must also update the checksums in the packet. Otherwise, the packet would be detected as containing errors and discarded by the next router. The NAT router must also remember which private IP address sent the packet in order to route the reply. Without remembering the translation address, the NAT router wouldn't know where to send the reply to on the private network.

NATs will also modify the packet data in some cases. For example, in the **File Transfer Protocol** (**FTP**), some connection information is sent as part of the packet's data. In these cases, the NAT router will look at the packet's data in order to know how to forward future incoming packets. IPv6 largely avoids the need for NAT, as it is possible (and common) for each device to have its own publicly-addressable address.

You may be wondering how a router knows whether a message is locally deliverable or whether it must be forwarded. This is done using a netmask, subnet mask, or CIDR.

Subnetting and CIDR

IP addresses can be split into parts. The most significant bits are used to identify the network or subnetwork, and the least significant bits are used to identify the specific device on the network.

This is similar to how your home address can be split into parts. Your home address includes a house number, a street name, and a city. The city is analogous to the network part, the street name could be the subnetwork part, and your house number is the device part.

IPv4 traditionally uses a mask notation to identify the IP address parts. For example, consider a router on the 10.0.0.0 network with a subnet mask of 255.255.255.0. This router can take any incoming packet and perform a bitwise AND operation with the subnet mask to determine whether the packet belongs on the local subnet or needs to be forwarded on. For example, this router receives a packet to be delivered to 10.0.0.105. It does a bitwise AND operation on this address with the subnet mask of 255.255.255.0, which produces 10.0.0.0. That matches the subnet of the router, so the traffic is local. If, instead, we consider a packet destined for 10.0.15.22, the result of the bitwise AND with the subnet mask is 10.0.15.0. This address doesn't match the subnet the router is on, and so it must be forwarded.

IPv6 uses CIDR. Networks and subnetworks are specified using the CIDR notation we described earlier. For example, if the IPv6 subnet is /112, then the router knows that any address that matches on the first 112 bits is on the local subnet.

So far, we've covered only routing with one sender and one receiver. While this is the most common situation, let's consider alternative cases too.

Multicast, broadcast, and anycast

When a packet is routed from one sender to one receiver, it uses **unicast** addressing. This is the simplest and most common type of addressing. All of the protocols we deal with in this book use unicast addressing.

Broadcast addressing allows a single sender to address a packet to all recipients simultaneously. It is typically used to deliver a packet to every receiver on an entire subnet.

If a broadcast is a one-to-all communication, then **multicast** is a one-to-many communication. Multicast involves some group management, and a message is addressed and delivered to members of a group.

Anycast addressed packets are used to deliver a message to one recipient when you don't care who that recipient is. This is useful if you have several servers that provide the same functionality, and you simply want one of them (you don't care which) to handle your request.

IPv4 and lower network levels support local broadcast addressing. IPv4 provides some optional (but commonly implemented) support for multicasting. IPv6 mandates multicasting support while providing additional features over IPv4's multicasting. Though IPv6 is not considered to broadcast, its multicasting functionality can essentially emulate it.

It's worth noting that these alternative addressing methods don't generally work over the broader internet. Imagine if one peer was able to broadcast a packet to every connected internet device. It would be a mess!

If you can use IP multicasting on your local network, though, it is worthwhile to implement it. Sending one IP level multicast conserves bandwidth compared to sending the same unicast message multiple times.

However, multicasting is often done at the application level. That is, when the application wants to deliver the same message to several recipients, it sends the message multiple times – once to each recipient. In `Chapter 3`, *An In-Depth Overview of TCP Connections*, we build a chat room. This chat room could be said to use application-level multicasting, but it does not take advantage of IP multicasting.

We've covered how messages are routed through a network. Now, let's see how a message knows which application is responsible for it once it arrives at a specific system.

Port numbers

An IP address alone isn't quite enough. We need port numbers. To return to the telephone analogy, if IP addresses are phone numbers, then port numbers are like phone extensions.

Generally, an IP address gets a packet routed to a specific system, but a port number is used to route the packet to a specific application on that system.

For example, on your system, you may be running multiple web browsers, an email client, and a video-conferencing client. When your computer receives a TCP segment or UDP datagram, your operating system looks at the destination port number in that packet. That port number is used to look up which application should handle it.

Port numbers are stored as unsigned 16-bit integers. This means that they are between 0 and 65,535 inclusive.

Some port numbers for common protocols are as follows:

Port Number		Protocol	
20, 21	TCP	File Transfer Protocol (FTP)	
22	TCP	Secure Shell (SSH)	Chapter 11, *Establishing SSH Connections with libssh*
23	TCP	Telnet	
25	TCP	Simple Mail Transfer Protocol (SMTP)	Chapter 8, *Making Your Program Send Email*
53	UDP	Domain Name System (DNS)	Chapter 5, *Hostname Resolution and DNS*
80	TCP	Hypertext Transfer Protocol (HTTP)	Chapter 6, *Building a Simple Web Client* Chapter 7, *Building a Simple Web Server*
110	TCP	Post Office Protocol, Version 3 (POP3)	
143	TCP	Internet Message Access Protocol (IMAP)	
194	TCP	Internet Relay Chat (IRC)	
443	TCP	HTTP over TLS/SSL (HTTPS)	Chapter 9, *Loading Secure Web Pages with HTTPS and OpenSSL* Chapter 10, *Implementing a Secure Web Server*
993	TCP	IMAP over TLS/SSL (IMAPS)	
995	TCP	POP3 over TLS/SSL (POP3S)	

Each of these listed port numbers is assigned by the **Internet Assigned Numbers Authority (IANA)**. They are responsible for the official assignments of port numbers for specific protocols. Unofficial port usage is very common for applications implementing custom protocols. In this case, the application should try to choose a port number that is not in common use to avoid conflict.

Clients and servers

In the telephone analogy, a call must be initiated first by one party. The initiating party dials the number for the receiving party, and the receiving party answers.

This is also a common paradigm in networking called the **client-server** model. In this model, a server listens for connections. The client, knowing the address and port number that the server is listening on, establishes the connection by sending the first packet.

For example, the web server at `example.com` listens on port `80` (HTTP) and port `443` (HTTPS). A web browser (client) must establish the connection by sending the first packet to the web server address and port.

Putting it together

A socket is one end-point of a communication link between systems. It's an abstraction in which your application can send and receive data over the network, in much the same way that your application can read and write to a file using a file handle.

An open socket is uniquely defined by a 5-tuple consisting of the following:

- Local IP address
- Local port
- Remote IP address
- Remote port
- Protocol (UDP or TCP)

This 5-tuple is important, as it is how your operating system knows which application is responsible for any packets received. For example, if you use two web browsers to establish two simultaneous connections to `example.com` on port `80`, then your operating system keeps the connections separate by looking at the local IP address, local port, remote IP address, remote port, and protocol. In this case, the local IP addresses, remote IP addresses, remote port (`80`), and protocol (TCP) are identical.

The deciding factor then is the local port (also called the **ephemeral port**), which will have been chosen to be different by the operating system for connection. This 5-tuple is also important to understand how NAT works. A private network may have many systems accessing the same outside resource, and the router NAT must store this five tuple for each connection in order to know how to route received packets back into the private network.

What's your address?

You can find your IP address using the `ipconfig` command on Windows, or the `ifconfig` command on Unix-based systems (such as Linux and macOS).

Using the `ipconfig` command from **Windows PowerShell** looks like this:

```
Windows PowerShell                                          —    □    ×
PS C:\Users\honp> ipconfig

Windows IP Configuration


Ethernet adapter Ethernet0:

    Connection-specific DNS Suffix  . : localdomain
    Link-local IPv6 Address . . . . . : fe80::cd70:e700:5486:fb1a%5
    IPv4 Address. . . . . . . . . . . : 192.168.182.133
    Subnet Mask . . . . . . . . . . . : 255.255.255.0
    Default Gateway . . . . . . . . . : 192.168.182.2

Tunnel adapter isatap.localdomain:

    Media State . . . . . . . . . . . : Media disconnected
    Connection-specific DNS Suffix  . : localdomain

Tunnel adapter Local Area Connection* 2:

    Connection-specific DNS Suffix  . :
    IPv6 Address. . . . . . . . . . . : 2001:0:9d38:6ab8:8ba:2c5a:5950:c03c
    Link-local IPv6 Address . . . . . : fe80::8ba:2c5a:5950:c03c%2
    Default Gateway . . . . . . . . . : ::
PS C:\Users\honp>
```

In this example, you can find that the IPv4 address is listed under `Ethernet adapter Ethernet0`. Your system may have more network adapters, and each will have its own IP address. We can tell that this computer is on a local network because the IP address, `192.168.182.133`, is in the private IP address range.

On Unix-based systems, we use either the `ifconfig` or `ip addr` commands. The `ifconfig` command is the old way and is now deprecated on some systems. The `ip addr` command is the new way, but not all systems support it yet.

Using the `ifconfig` command from a macOS terminal looks like this:

```
                          ⌂ bob — bash — 80×20
Last login: Mon Sep 17 19:22:53 on ttys000
m1:~ honp$ ifconfig
lo0: flags=8049<UP,LOOPBACK,RUNNING,MULTICAST> mtu 16384
        options=3<RXCSUM,TXCSUM>
        inet6 fe80::1%lo0 prefixlen 64 scopeid 0x1
        inet 127.0.0.1 netmask 0xff000000
        inet6 ::1 prefixlen 128
gif0: flags=8010<POINTOPOINT,MULTICAST> mtu 1280
stf0: flags=0<> mtu 1280
en0: flags=8863<UP,BROADCAST,SMART,RUNNING,SIMPLEX,MULTICAST> mtu 1500
        options=b<RXCSUM,TXCSUM,VLAN_HWTAGGING>
        ether 00:0c:29:59:17:6f
        inet6 fe80::20c:29ff:fe59:176f%en0 prefixlen 64 scopeid 0x4
        inet 192.168.182.128 netmask 0xffffff00 broadcast 192.168.182.255
        media: autoselect (1000baseT <full-duplex>)
        status: active
m1:~ honp$
```

The IPv4 address is listed next to `inet`. In this case, we can see that it's `192.168.182.128`. Again, we see that this computer is on a local network because of the IP address range. The same adapter has an IPv6 address listed next to `inet6`.

The following screenshot shows using the `ip addr` command on Ubuntu Linux:

```
                                honp@ubby18: ~
honp@ubby18:~$ ip addr
1: lo: <LOOPBACK,UP,LOWER_UP> mtu 65536 qdisc noqueue state UNKNOWN group default qlen 1000
    link/loopback 00:00:00:00:00:00 brd 00:00:00:00:00:00
    inet 127.0.0.1/8 scope host lo
       valid_lft forever preferred_lft forever
    inet6 ::1/128 scope host
       valid_lft forever preferred_lft forever
2: ens33: <BROADCAST,MULTICAST,UP,LOWER_UP> mtu 1500 qdisc fq_codel state UP group default qlen 1000
    link/ether 00:0c:29:74:ba:ce brd ff:ff:ff:ff:ff:ff
    inet 192.168.182.145/24 brd 192.168.182.255 scope global dynamic noprefixroute ens33
       valid_lft 1515sec preferred_lft 1515sec
    inet6 fe80::df60:954e:211:7ff0/64 scope link noprefixroute
       valid_lft forever preferred_lft forever
honp@ubby18:~$
```

The preceding screenshot shows the local IPv4 address as `192.168.182.145`. We can also see that the link-local IPv6 address is `fe80::df60:954e:211:7ff0`.

These commands, `ifconfig`, `ip addr`, and `ipconfig`, show the IP address or addresses for each adapter on your computer. You may have several. If you are on a local network, the IP addresses you see will be your local private network IP addresses.

If you are behind a NAT, there is often no good way to know your public IP address. Usually, the only resort is to contact an internet server that provides an API that informs you of your IP address.

A few free and public APIs for this are as follows:

- `http://api.ipify.org/`
- `http://helloacm.com/api/what-is-my-ip-address/`
- `http://icanhazip.com/`
- `http://ifconfig.me/ip`

You can test out these APIs in a web browser:

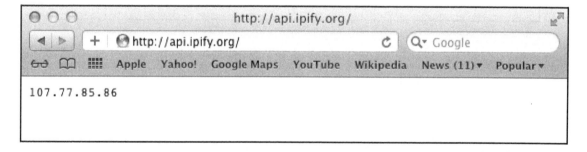

Each of these listed web pages should return your public IP address and not much else. These sites are useful for when you need to determine your public IP address from behind an NAT programmatically. We look at writing a small HTTP client capable of downloading these web pages and others in Chapter 6, *Building a Simple Web Client*.

Now that we've seen the built-in utilities for determining our local IP addresses, let's next look at how to accomplish this from C.

Listing network adapters from C

Sometimes, it is useful for your C programs to know what your local address is. For most of this book, we are able to write code that works both on Windows and Unix-based (Linux and macOS) systems. However, the API for listing local addresses is very different between systems. For this reason, we split this program into two: one for Windows and one for Unix-based systems.

We will address the Windows case first.

Listing network adapters on Windows

The Windows networking API is called **Winsock**, and we will go into much more detail about it in the next chapter.

Whenever we are using Winsock, the first thing we must do is initialize it. This is done with a call to WSAStartup(). Here is a small C program, win_init.c, showing the initialization and cleanup of Winsock:

```
/*win_init.c*/

#include <stdio.h>
#include <winsock2.h>
#pragma comment(lib, "ws2_32.lib")

int main() {
    WSADATA d;

    if (WSAStartup(MAKEWORD(2, 2), &d)) {
        printf("Failed to initialize.\n");
        return -1;
    }

    WSACleanup();
    printf("Ok.\n");
    return 0;
}
```

The WSAStartup() function is called with the requested version, Winsock 2.2 in this case, and a WSADATA structure. The WSADATA structure will be filled in by WSAStartup() with details about the Windows Sockets implementation. The WSAStartup() function returns 0 upon success, and non-zero upon failure.

When a Winsock program is finished, it should call `WSACleanup()`.

If you are using Microsoft Visual C as your compiler, then `#pragma comment(lib, "ws2_32.lib")` tells Microsoft Visual C to link the executable with the Winsock library, `ws2_32.lib`.

If you are using MinGW as your compiler, the pragma is ignored. You need to explicitly tell the compiler to link in the library by adding the command-line option, `-lws2_32`. For example, you can compile this program using MinGW with the following command:

```
gcc win_init.c -o win_init.exe -lws2_32
```

We will cover Winsock initialization and usage in more detail in `Chapter 2`, *Getting to Grips with Socket APIs*.

Now that we know how to initialize Winsock, we will begin work on the complete program to list network adapters on Windows. Please refer to the `win_list.c` file to follow along.

To begin with, we need to define `_WIN32_WINNT` and include the needed headers:

```
/*win_list.c*/

#ifndef _WIN32_WINNT
#define _WIN32_WINNT 0x0600
#endif

#include <winsock2.h>
#include <iphlpapi.h>
#include <ws2tcpip.h>
#include <stdio.h>
#include <stdlib.h>
```

The `_WIN32_WINNT` macro must be defined first so that the proper version of the Windows headers are included. `winsock2.h`, `iphlpapi.h`, and `ws2tcpip.h` are the Windows headers we need in order to list network adapters. We need `stdio.h` for the `printf()` function and `stdlib.h` for memory allocation.

Next, we include the following pragmas to tell Microsoft Visual C which libraries must be linked with the executable:

```
/*win_list.c continued*/

#pragma comment(lib, "ws2_32.lib")
#pragma comment(lib, "iphlpapi.lib")
```

If you're compiling with MinGW, these lines will have no effect. You will need to link to these libraries explicitly on the command line, for example, `gcc win_list.c -o win_list.exe -liphlpapi -lws2_32`.

We then enter the `main()` function and initialize Winsock 2.2 using `WSAStartup()` as described earlier. We check its return value to detect any errors:

```
/*win_list.c continued*/

int main() {

    WSADATA d;
    if (WSAStartup(MAKEWORD(2, 2), &d)) {
        printf("Failed to initialize.\n");
        return -1;
    }
```

Next, we allocate memory for the adapters, and we request the adapters' addresses from Windows using the `GetAdapterAddresses()` function:

```
/*win_list.c continued*/

    DWORD asize = 20000;
    PIP_ADAPTER_ADDRESSES adapters;
    do {
        adapters = (PIP_ADAPTER_ADDRESSES)malloc(asize);

        if (!adapters) {
            printf("Couldn't allocate %ld bytes for adapters.\n", asize);
            WSACleanup();
            return -1;
        }

        int r = GetAdaptersAddresses(AF_UNSPEC, GAA_FLAG_INCLUDE_PREFIX, 0,
                adapters, &asize);
        if (r == ERROR_BUFFER_OVERFLOW) {
            printf("GetAdaptersAddresses wants %ld bytes.\n", asize);
            free(adapters);
        } else if (r == ERROR_SUCCESS) {
            break;
        } else {
            printf("Error from GetAdaptersAddresses: %d\n", r);
            free(adapters);
            WSACleanup();
            return -1;
        }
    } while (!adapters);
```

The `asize` variable will store the size of our adapters' address buffer. To begin with, we set it to `20000` and allocate 20,000 bytes to `adapters` using the `malloc()` function. The `malloc()` function will return 0 on failure, so we test for that and display an error message if allocation failed.

Next, we call `GetAdapterAddresses()`. The first parameter, `AF_UNSPEC`, tells Windows that we want both IPv4 and IPv6 addresses. You can pass in `AF_INET` or `AF_INET6` to request only IPv4 or only IPv6 addresses. The second parameter, `GAA_FLAG_INCLUDE_PREFIX`, is required to request a list of addresses. The next parameter is reserved and should be passed in as 0 or `NULL`. Finally, we pass in our buffer, `adapters`, and a pointer to its size, `asize`.

If our buffer is not big enough to store all of the addresses, then `GetAdapterAddresses()` returns `ERROR_BUFFER_OVERFLOW` and sets `asize` to the required buffer size. In this case, we free our `adapters` buffer and try the call again with a larger buffer.

On success, `GetAdapterAddresses()` returns `ERROR_SUCCESS`, in which case, we break from the loop and continue. Any other return value is an error.

When `GetAdapterAddresses()` returns successfully, it will have written a linked list into `adapters` with each adapter's address information. Our next step is to loop through this linked list and print information for each adapter and address:

```
/*win_list.c continued*/

    PIP_ADAPTER_ADDRESSES adapter = adapters;
    while (adapter) {
        printf("\nAdapter name: %S\n", adapter->FriendlyName);

        PIP_ADAPTER_UNICAST_ADDRESS address = adapter->FirstUnicastAddress;
        while (address) {
            printf("\t%s",
                    address->Address.lpSockaddr->sa_family == AF_INET ?
                    "IPv4" : "IPv6");

            char ap[100];

            getnameinfo(address->Address.lpSockaddr,
                    address->Address.iSockaddrLength,
                    ap, sizeof(ap), 0, 0, NI_NUMERICHOST);
            printf("\t%s\n", ap);

            address = address->Next;
        }
```

```
        adapter = adapter->Next;
    }
```

We first define a new variable, `adapter`, which we use to walk through the linked list of adapters. The first adapter is at the beginning of `adapters`, so we initially set `adapter` to `adapters`. At the end of each loop, we set `adapter = adapter->Next;` to get the next adapter. The loop aborts when `adapter` is 0, which means we've reached the end of the list.

We get the adapter name from `adapter->FriendlyName`, which we then print using `printf()`.

The first address for each adapter is in `adapter->FirstUnicastAddress`. We define a second pointer, `address`, and set it to this address. Addresses are also stored as a linked list, so we begin an inner loop that walks through the addresses.

The `address->Address.lpSockaddr->sa_family` variable stores the address family type. If it is set to `AF_INET`, then we know this is an IPv4 address. Otherwise, we assume it is an IPv6 address (in which case the family is `AF_INET6`).

Next, we allocate a buffer, `ap`, to store the text representation of the address. The `getnameinfo()` function is called to convert the address into a standard notation address. We'll cover more about `getnameinfo()` in the next chapter.

Finally, we can print the address from our buffer, `ap`, using `printf()`.

We finish the program by freeing the allocated memory and calling `WSACleanup()`:

```
/*win_list.c continued*/

    free(adapters);
    WSACleanup();
    return 0;
}
```

On Windows, using MinGW, you can compile and run the program with the following:

```
gcc win_list.c -o win_list.exe -liphlpapi -lws2_32
win_list
```

It should list each of your adapter's names and addresses.

Now that we can list local IP addresses on Windows, let's consider the same task for Unix-based systems.

Listing network adapters on Linux and macOS

Listing local network addresses is somewhat easier on a Unix-based system, compared to Windows. Load up `unix_list.c` to follow along.

To begin with, we include the necessary system headers:

```
/*unix_list.c*/

#include <sys/socket.h>
#include <netdb.h>
#include <ifaddrs.h>
#include <stdio.h>
#include <stdlib.h>
```

We then enter the `main` function:

```
/*unix_list.c continued*/

int main() {

    struct ifaddrs *addresses;

    if (getifaddrs(&addresses) == -1) {
        printf("getifaddrs call failed\n");
        return -1;
    }
```

We declare a variable, `addresses`, which stores the addresses. A call to the `getifaddrs()` function allocates memory and fills in a linked list of addresses. This function returns 0 on success or -1 on failure.

Next, we use a new pointer, `address`, to walk through the linked list of addresses. After considering each address, we set `address = address->ifa_next` to get the next address. We stop the loop when `address == 0`, which happens at the end of the linked list:

```
/*unix_list.c continued*/

    struct ifaddrs *address = addresses;
    while(address) {
        int family = address->ifa_addr->sa_family;
        if (family == AF_INET || family == AF_INET6) {

            printf("%s\t", address->ifa_name);
            printf("%s\t", family == AF_INET ? "IPv4" : "IPv6");

            char ap[100];
```

```
        const int family_size = family == AF_INET ?
            sizeof(struct sockaddr_in) : sizeof(struct sockaddr_in6);
        getnameinfo(address->ifa_addr,
                family_size, ap, sizeof(ap), 0, 0, NI_NUMERICHOST);
        printf("\t%s\n", ap);


    }
    address = address->ifa_next;
}
```

For each address, we identify the address family. We are interested in AF_INET (IPv4 addresses) and AF_INET6 (IPv6 addresses). The getifaddrs() function can return other types, so we skip those.

For each address, we then continue to print its adapter name and its address type, IPv4 or IPv6.

We then define a buffer, ap, to store the textual address. A call to the getnameinfo() function fills in this buffer, which we can then print. We cover the getnameinfo() function in more detail in the next chapter, Chapter 2, *Getting to Grips with Socket APIs*.

Finally, we free the memory allocated by getifaddrs() and we have finished:

```
/*unix_list.c continued*/

    freeifaddrs(addresses);
    return 0;
}
```

On Linux and macOS, you can compile and run this program with the following:

```
gcc unix_list.c -o unix_list
./unix_list
```

It should list each of your adapter's names and addresses.

Summary

In this chapter, we looked briefly at how internet traffic is routed. We learned that there are two Internet Protocol versions, IPv4 and IPv6. IPv4 has a limited number of addresses, and these addresses are running out. One of IPv6's main advantages is that it has enough address space for every system to have its own unique publicly-routable address. The limited address space of IPv4 is largely mitigated by network address translation performed by routers. We also looked at how to detect your local IP address using both utilities and APIs provided by the operating system.

We saw that the APIs provided for listing local IP addresses differ quite a bit between Windows and Unix-based operating systems. In future chapters, we will see that most other networking functions are similar between operating systems, and we can write one portable program that works between operating systems.

It's OK if you didn't pick up all of the details in this chapter. Most of this information is a helpful background, but it's not essential to most network application programming. Details such as network address translation are handled by the network, and these details will not usually need to be explicitly addressed by your programs.

In the next chapter, we will reinforce the ideas covered here by introducing socket-programming APIs.

Questions

Try these questions to test your knowledge from this chapter:

1. What are the key differences between IPv4 and IPv6?
2. Are the IP addresses given by the `ipconfig` and `ifconfig` commands the same IP addresses that a remote web server sees if you connect to it?
3. What is the IPv4 loopback address?
4. What is the IPv6 loopback address?
5. How are domain names (for example, `example.com`) resolved into IP addresses?
6. How can you find your public IP address?
7. How does an operating system know which application is responsible for an incoming packet?

The answers are in `Appendix A`, *Answers to Questions*.

2

Getting to Grips with Socket APIs

In this chapter, we will begin to really start working with network programming. We will introduce the concept of sockets, and explain a bit of the history behind them. We will cover the important differences between the socket APIs provided by Windows and Unix-like operating systems, and we will review the common functions that are used in socket programming. This chapter ends with a concrete example of turning a simple console program into a networked program you can access through your web browser.

The following topics are covered in this chapter:

- What are sockets?
- Which header files are used with socket programming?
- How to compile a socket program on Windows, Linux, and macOS
- Connection-oriented and connectionless sockets
- TCP and UDP protocols
- Common socket functions
- Building a simple console program into a web server

Technical requirements

The example programs in this chapter can be compiled with any modern C compiler. We recommend MinGW on Windows and GCC on Linux and macOS. See Appendix B, *Setting Up Your C Compiler On Windows*, Appendix C, *Setting Up Your C Compiler On Linux*, and Appendix D, *Setting Up Your C Compiler On macOS*, for compiler setup.

The code for this book can be found here: https://github.com/codeplea/Hands-On-Network-Programming-with-C.

From the command line, you can download the code for this chapter with the following command:

```
git clone https://github.com/codeplea/Hands-On-Network-Programming-with-C
cd Hands-On-Network-Programming-with-C/chap02
```

Each example program in this chapter is standalone, and each example runs on Windows, Linux, and macOS. When compiling for Windows, keep in mind that most of the example programs require linking with the Winsock library.

This is accomplished by passing the `-lws2_32` option to `gcc`. We provide the exact commands needed to compile each example as they are introduced.

What are sockets?

A socket is one endpoint of a communication link between systems. Your application sends and receives all of its network data through a socket.

There are a few different socket **application programming interfaces** (**APIs**). The first were Berkeley sockets, which were released in 1983 with 4.3BSD Unix. The Berkeley socket API was widely successful and quickly evolved into a de facto standard. From there, it was adopted as a POSIX standard with little modification. The terms Berkeley sockets, BSD sockets, Unix sockets, and **Portable Operating System Interface** (**POSIX**) sockets are often used interchangeably.

If you're using Linux or macOS, then your operating system provides a proper implementation of Berkeley sockets.

Windows' socket API is called **Winsock**. It was created to be largely compatible with Berkeley sockets. In this book, we strive to create cross-platform code that is valid for both Berkeley sockets and Winsock.

Historically, sockets were used for **inter-process communication** (**IPC**) as well as various network protocols. In this book, we use sockets only for communication with TCP and UDP.

Before we can start using sockets, we need to do a bit of setup. Let's dive right in!

Socket setup

Before we can use the socket API, we need to include the socket API header files. These files vary depending on whether we are using Berkeley sockets or Winsock. Additionally, Winsock requires initialization before use. It also requires that a cleanup function is called when we are finished. These initialization and cleanup steps are not used with Berkeley sockets.

We will use the C preprocessor to run the proper code on Windows compared to Berkeley socket systems. By using the preprocessor statement, `#if defined(_WIN32)`, we can include code in our program that will only be compiled on Windows.

Here is a complete program that includes the needed socket API headers for each platform and properly initializes Winsock on Windows:

```
/*sock_init.c*/

#if defined(_WIN32)
#ifndef _WIN32_WINNT
#define _WIN32_WINNT 0x0600
#endif
#include <winsock2.h>
#include <ws2tcpip.h>
#pragma comment(lib, "ws2_32.lib")

#else
#include <sys/types.h>
#include <sys/socket.h>
#include <netinet/in.h>
#include <arpa/inet.h>
#include <netdb.h>
#include <unistd.h>
#include <errno.h>

#endif

#include <stdio.h>

int main() {

#if defined(_WIN32)
    WSADATA d;
    if (WSAStartup(MAKEWORD(2, 2), &d)) {
        fprintf(stderr, "Failed to initialize.\n");
        return 1;
    }
```

```
    #endif

        printf("Ready to use socket API.\n");

#if defined(_WIN32)
        WSACleanup();
#endif

        return 0;
}
```

The first part includes `winsock.h` and `ws2tcpip.h` on Windows. `_WIN32_WINNT` must be defined for the Winsock headers to provide all the functions we need. We also include the `#pragma comment(lib, "ws2_32.lib")` pragma statement. This tells the Microsoft Visual C compiler to link your program against the Winsock library, `ws2_32.lib`. If you're using MinGW as your compiler, then `#pragma` is ignored. In this case, you need to tell the compiler to link in `ws2_32.lib` on the command line using the `-lws2_32` option.

If the program is not compiled on Windows, then the section after `#else` will compile. This section includes the various Berkeley socket API headers and other headers we need on these platforms.

In the `main()` function, we call `WSAStartup()` on Windows to initialize Winsock. The `MAKEWORD` macro allows us to request Winsock version 2.2. If our program is unable to initialize Winsock, it prints an error message and aborts.

When using Berkeley sockets, no special initialization is needed, and the socket API is always ready to use.

Before our program finishes, `WSACleanup()` is called if we're compiling for Winsock on Windows. This function allows the Windows operating system to do additional cleanup.

Compiling and running this program on Linux or macOS is done with the following command:

```
gcc sock_init.c -o sock_init
./sock_init
```

Compiling on Windows using MinGW can be done with the following command:

```
gcc sock_init.c -o sock_init.exe -lws2_32
sock_init.exe
```

Notice that the `-lws2_32` flag is needed with MinGW to tell the compiler to link in the Winsock library, `ws2_32.lib`.

Now that we've done the necessary setup to begin using the socket APIs, let's take a closer look at what we will be using these sockets for.

Two types of sockets

Sockets come in two basic types—**connection-oriented** and **connectionless**. These terms refer to types of protocols. Beginners sometimes get confused with the term **connectionless**. Of course, two systems communicating over a network are in some sense connected. Keep in mind that these terms are used with special meanings, which we will cover shortly, and should not imply that some protocols manage to send data without a connection.

The two protocols that are used today are **Transmission Control Protocol** (**TCP**) and **User Datagram Protocol** (**UDP**). TCP is a connection-oriented protocol, and UDP is a connectionless protocol.

The socket APIs also support other less-common or outdated protocols, which we do not cover in this book.

In a connectionless protocol, such as UDP, each data packet is addressed individually. From the protocol's perspective, each data packet is completely independent and unrelated to any packets coming before or after it.

A good analogy for UDP is **postcards**. When you send a postcard, there is no guarantee that it will arrive. There is also no way to know if it did arrive. If you send many postcards at once, there is no way to predict what order they will arrive in. It is entirely possible that the first postcard you send gets delayed and arrives weeks after the last postcard was sent.

With UDP, these same caveats apply. UDP makes no guarantee that a packet will arrive. UDP doesn't generally provide a method to know if a packet did not arrive, and UDP does not guarantee that the packets will arrive in the same order they were sent. As you can see, UDP is no more reliable than postcards. In fact, you may consider it less reliable, because with UDP, it is possible that a single packet may arrive twice!

If you need reliable communication, you may be tempted to develop a scheme where you number each packet that's sent. For the first packet sent, you number it one, the second packet sent is numbered two, and so on. You could also request that the receiver send an acknowledgment for each packet. When the receiver gets packet one, it sends a return message, **packet one received**. In this way, the receiver can be sure that received packets are in the proper order. If the same packet arrives twice, the receiver can just ignore the redundant copy. If a packet isn't received at all, the sender knows from the missing acknowledgment and can resend it.

This scheme is essentially what connection-oriented protocols, such as TCP, do. TCP guarantees that data arrives in the same order it is sent. It prevents duplicate data from arriving twice, and it retries sending missing data. It also provides additional features such as notifications when a connection is terminated and algorithms to mitigate network congestion. Furthermore, TCP implements these features with an efficiency that is not achievable by piggybacking a custom reliability scheme on top of UDP.

For these reasons, TCP is used by many protocols. HTTP (for severing web pages), FTP (for transferring files), SSH (for remote administration), and SMTP (for delivering email) all use TCP. We will cover HTTP, SSH, and SMTP in the coming chapters.

UDP is used by DNS (for resolving domain names). It is suitable for this purpose because an entire request and response can fit in a single packet.

UDP is also commonly used in real-time applications, such as audio streaming, video streaming, and multiplayer video games. In real-time applications, there is often no reason to retry sending dropped packets, so TCP's guarantees are unnecessary. For example, if you are streaming live video and a few packets get dropped, the video simply resumes when the next packet arrives. There is no reason to resend (or even detect) the dropped packet, as the video has already progressed past that point.

UDP also has the advantage in cases where you want to send a message without expecting a response from the other end. This makes it useful when using IP broadcast or multicast. TCP, on the other hand, requires bidirectional communication to provide its guarantees, and TCP does not work with IP multicast or broadcast.

If the guarantees that TCP provides are not needed, then UDP can achieve greater efficiency. This is because TCP adds some additional overhead by numbering packets. TCP must also delay packets that arrive out of order, which can cause unnecessary delays in real-time applications. If you do need the guarantees provided by TCP, however, it is almost always preferable to use TCP instead of trying to add those mechanisms to UDP.

Now that we have an idea of the communication models we use sockets for, let's look at the actual functions that are used in socket programming.

Socket functions

The socket APIs provide many functions for use in network programming. Here are the common socket functions that we use in this book:

- `socket()` creates and initializes a new socket.
- `bind()` associates a socket with a particular local IP address and port number.
- `listen()` is used on the server to cause a TCP socket to listen for new connections.
- `connect()` is used on the client to set the remote address and port. In the case of TCP, it also establishes a connection.
- `accept()` is used on the server to create a new socket for an incoming TCP connection.
- `send()` and `recv()` are used to send and receive data with a socket.
- `sendto()` and `recvfrom()` are used to send and receive data from sockets without a bound remote address.
- `close()` (Berkeley sockets) and `closesocket()` (Winsock sockets) are used to close a socket. In the case of TCP, this also terminates the connection.
- `shutdown()` is used to close one side of a TCP connection. It is useful to ensure an orderly connection teardown.
- `select()` is used to wait for an event on one or more sockets.
- `getnameinfo()` and `getaddrinfo()` provide a protocol-independent manner of working with hostnames and addresses.
- `setsockopt()` is used to change some socket options.
- `fcntl()` (Berkeley sockets) and `ioctlsocket()` (Winsock sockets) are also used to get and set some socket options.

You may see some Berkeley socket networking programs using `read()` and `write()`. These functions don't port to Winsock, so we prefer `send()` and `recv()` here. Some other common functions that are used with Berkeley sockets are `poll()` and `dup()`. We will avoid these in order to keep our programs portable.

Other differences between Berkeley sockets and Winsock sockets are addressed later in this chapter.

Now that we have an idea of the functions involved, let's consider program design and flow next.

Anatomy of a socket program

As we mentioned in Chapter 1, *An Introduction to Networks and Protocols*, network programming is usually done using a client-server paradigm. In this paradigm, a server listens for new connections at a published address. The client, knowing the server's address, is the one to establish the connection initially. Once the connection is established, the client and the server can both send and receive data. This can continue until either the client or the server terminates the connection.

A traditional client-server model usually implies different behaviors for the client and server. The way web browsing works, for example, is that the server resides at a known address, waiting for connections. A client (web browser) establishes a connection and sends a request that includes which web page or resource it wants to download. The server then checks that it knows what to do with this request and responds appropriately (by sending the web page).

An alternative paradigm is the peer-to-peer model. For example, this model is used by the BitTorrent protocol. In the peer-to-peer model, each peer has essentially the same responsibilities. While a web server is optimized to send requested data from the server to the client, a peer-to-peer protocol is balanced in that data is exchanged somewhat evenly between peers. However, even in the peer-to-peer model, the underlying sockets that are using TCP or UDP aren't created equal. That is, for each peer-to-peer connection, one peer was listening and the other connecting. BitTorrent works by having a central server (called a **tracker**) that stores a list of peer IP addresses. Each of the peers on that list has agreed to behave like a server and listen for new connections. When a new peer wants to join the swarm, it requests a list of peers from the central server, and then tries to establish a connection to peers on that list while simultaneously listening for new connections from other peers. In summary, a peer-to-peer protocol doesn't so much replace the client-server model; it is just expected that each peer be a client and a server both.

Another common protocol that pushes the boundary of the client-server paradigm is FTP. The FTP server listens for connections until the FTP client connects. After the initial connection, the FTP client issues commands to the server. If the FTP client requests a file from the server, the server will attempt to establish a new connection to the FTP client to transfer the file over. So, for this reason, the FTP client first establishes a connection as a TCP client, but later accepts connections like a TCP server.

Network programs can usually be described as one of four types—a TCP server, a TCP client, a UDP server, or a UDP client. Some protocols call for a program to implement two, or even all four types, but it is useful for us to consider each of the four types separately.

TCP program flow

A TCP client program must first know the TCP server's address. This is often input by a user. In the case of a web browser, the server address is either input directly by the user into the address bar, or is known from the user clicking on a link. The TCP client takes this address (for example, http://example.com) and uses the getaddrinfo() function to resolve it into a struct addrinfo structure. The client then creates a socket using a call to socket(). The client then establishes the new TCP connection by calling connect(). At this point, the client can freely exchange data using send() and recv().

A TCP server listens for connections at a particular port number on a particular interface. The program must first initialize a struct addrinfo structure with the proper listening IP address and port number. The getaddrinfo() function is helpful so that you can do this in an IPv4/IPv6 independent way. The server then creates the socket with a call to socket(). The socket must be bound to the listening IP address and port. This is accomplished with a call to bind().

The server program then calls listen(), which puts the socket in a state where it listens for new connections. The server can then call accept(), which will wait until a client establishes a connection to the server. When the new connection has been established, accept() returns a new socket. This new socket can be used to exchange data with the client using send() and recv(). Meanwhile, the first socket remains listening for new connections, and repeated calls to accept() allow the server to handle multiple clients.

Graphically, the program flow of a TCP client and server looks like this:

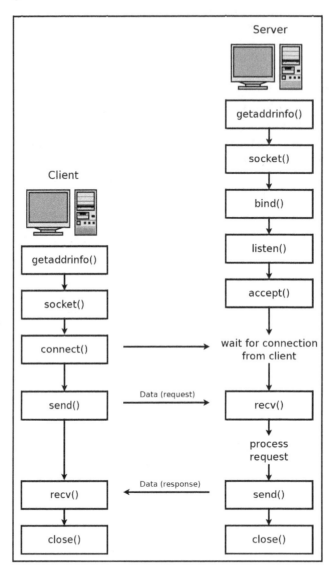

The program flow given here should serve as a good example of how basic client-server TCP programs interact. That said, considerable variation on this basic program flow is possible. There is also no rule about which side calls `send()` or `recv()` first, or how many times. Both sides could call `send()` as soon as the connection is established.

Also, note that the TCP client could call `bind()` before `connect()` if it is particular about which network interface is being used to connect with. This is sometimes important on servers that have multiple network interfaces. It's often not important for general purpose software.

Many other variations of TCP operation are possible too, and we will look at some in `Chapter 3`, *An In-Depth Overview of TCP Connections.*

UDP program flow

A UDP client must know the address of the remote UDP peer in order to send the first packet. The UDP client uses the `getaddrinfo()` function to resolve the address into a `struct addrinfo` structure. Once this is done, the client creates a socket of the proper type. The client can then call `sendto()` on the socket to send the first packet. The client can continue to call `sendto()` and `recvfrom()` on the socket to send and receive additional packets. Note that the client must send the first packet with `sendto()`. The UDP client cannot receive data first, as the remote peer would have no way of knowing where to send data without it first receiving data from the client. This is different from TCP, where a connection is first established with a handshake. In TCP, either the client or server can send the first application data.

A UDP server listens for connections from a UDP client. This server should initialize `struct addrinfo` structure with the proper listening IP address and port number. The `getaddrinfo()` function can be used to do this in a protocol-independent way. The server then creates a new socket with `socket()` and binds it to the listening IP address and port number using `bind()`. At this point, the server can call `recvfrom()`, which causes it to block until it receives data from a UDP client. After the first data is received, the server can reply with `sendto()` or listen for more data (from the first client or any new client) with `recvfrom()`.

Graphically, the program flow of a UDP client and server looks like this:

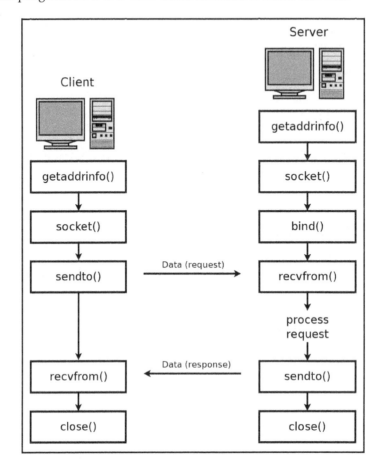

We cover some variations of this example program flow in `Chapter 4`, *Establishing UDP Connections*.

We're almost ready to begin implementing our first networked program, but before we begin, we should take care of some cross-platform concerns. Let's work on this now.

Berkeley sockets versus Winsock sockets

As we stated earlier, Winsock sockets were modeled on Berkeley sockets. Therefore, there are many similarities between them. However, there are also many differences we need to be aware of.

In this book, we will try to create each program so that it can run on both Windows and Unix-based operating systems. This is made much easier by defining a few C macros to help us with this.

Header files

As we mentioned earlier, the needed header files differ between implementations. We've already seen how these header file discrepancies can be easily overcome with a preprocessor statement.

Socket data type

In UNIX, a socket descriptor is represented by a standard file descriptor. This means you can use any of the standard UNIX file I/O functions on sockets. This isn't true on Windows, so we simply avoid these functions to maintain portability.

Additionally, in UNIX, all file descriptors (and therefore socket descriptors) are small, non-negative integers. In Windows, a socket handle can be anything. Furthermore, in UNIX, the `socket()` function returns an `int`, whereas in Windows it returns a `SOCKET`. `SOCKET` is a `typedef` for an `unsigned int` in the Winsock headers. As a workaround, I find it useful to either `typedef int SOCKET` or `#define SOCKET int` on non-Windows platforms. That way, you can store a socket descriptor as a `SOCKET` type on all platforms:

```
#if !defined(_WIN32)
#define SOCKET int
#endif
```

Invalid sockets

On Windows, `socket()` returns `INVALID_SOCKET` if it fails. On Unix, `socket()` returns a negative number on failure. This is particularly problematic as the Windows `SOCKET` type is unsigned. I find it useful to define a macro to indicate if a socket descriptor is valid or not:

```
#if defined(_WIN32)
#define ISVALIDSOCKET(s) ((s) != INVALID_SOCKET)
#else
#define ISVALIDSOCKET(s) ((s) >= 0)
#endif
```

Closing sockets

All sockets on Unix systems are also standard file descriptors. For this reason, sockets on Unix systems can be closed using the standard `close()` function. On Windows, a special close function is used instead—`closesocket()`. It's useful to abstract out this difference with a macro:

```
#if defined(_WIN32)
#define CLOSESOCKET(s) closesocket(s)
#else
#define CLOSESOCKET(s) close(s)
#endif
```

Error handling

When a socket function, such as `socket()`, `bind()`, `accept()`, and so on, has an error on a Unix platform, the error number gets stored in the thread-global `errno` variable. On Windows, the error number can be retrieved by calling `WSAGetLastError()` instead. Again, we can abstract out this difference using a macro:

```
#if defined(_WIN32)
#define GETSOCKETERRNO() (WSAGetLastError())
#else
#define GETSOCKETERRNO() (errno)
#endif
```

In addition to obtaining an error code, it is often useful to retrieve a text description of the error condition. Please refer to `Chapter 13`, *Socket Programming Tips and Pitfalls*, for a technique for this.

With these helper macros out of the way, let's dive into our first real socket program.

Our first program

Now that we have a basic idea of socket APIs and the structure of networked programs, we are ready to begin our first program. By building an actual real-world program, we will learn the useful details of how socket programming actually works.

As an example task, we are going to build a web server that tells you what time it is right now. This could be a useful resource for anybody with a smartphone or web browser that needs to know what time it is right now. They can simply navigate to our web page and find out. This is a good first example because it does something useful but still trivial enough that it won't distract from what we are trying to learn—network programming.

A motivating example

Before we begin the networked program, it is useful to solve our problem with a simple console program first. In general, it is a good idea to work out your program's functionality locally before adding in networked features.

The local, console version of our time-telling program is as follows:

```
/*time_console.c*/

#include <stdio.h>
#include <time.h>

int main()
{
    time_t timer;
    time(&timer);

    printf ("Local time is: %s", ctime(&timer));

    return 0;
}
```

You can compile and run it like this:

```
$ gcc time_console.c -o time_console
$ ./time_console
Local time is: Fri Oct 19 08:42:05 2018
```

The program works by getting the time with the built-in C time() function. It then converts it into a string with the ctime() function.

Making it networked

Now that we've worked out our program's functionality, we can begin on the networked version of the same program.

To begin with, we include the needed headers:

```
/*time_server.c*/

#if defined(_WIN32)
#ifndef _WIN32_WINNT
#define _WIN32_WINNT 0x0600
#endif
#include <winsock2.h>
#include <ws2tcpip.h>
#pragma comment(lib, "ws2_32.lib")

#else
#include <sys/types.h>
#include <sys/socket.h>
#include <netinet/in.h>
#include <arpa/inet.h>
#include <netdb.h>
#include <unistd.h>
#include <errno.h>

#endif
```

As we discussed earlier, this detects if the compiler is running on Windows or not and includes the proper headers for the platform it is running on.

We also define some macros, which abstract out some of the difference between the Berkeley socket and Winsock APIs:

```
/*time_server.c continued*/

#if defined(_WIN32)
#define ISVALIDSOCKET(s) ((s) != INVALID_SOCKET)
#define CLOSESOCKET(s) closesocket(s)
#define GETSOCKETERRNO() (WSAGetLastError())

#else
#define ISVALIDSOCKET(s) ((s) >= 0)
#define CLOSESOCKET(s) close(s)
#define SOCKET int
#define GETSOCKETERRNO() (errno)
#endif
```

We need a couple of standard C headers, hopefully for obvious reasons:

```
/*time_server.c continued*/

#include <stdio.h>
#include <string.h>
#include <time.h>
```

Now, we are ready to begin the main() function. The first thing the main() function will do is initialize Winsock if we are compiling on Windows:

```
/*time_server.c continued*/

int main() {

#if defined(_WIN32)
    WSADATA d;
    if (WSAStartup(MAKEWORD(2, 2), &d)) {
        fprintf(stderr, "Failed to initialize.\n");
        return 1;
    }
#endif
```

We must now figure out the local address that our web server should bind to:

```
/*time_server.c continued*/

    printf("Configuring local address...\n");
    struct addrinfo hints;
    memset(&hints, 0, sizeof(hints));
    hints.ai_family = AF_INET;
    hints.ai_socktype = SOCK_STREAM;
    hints.ai_flags = AI_PASSIVE;

    struct addrinfo *bind_address;
    getaddrinfo(0, "8080", &hints, &bind_address);
```

We use getaddrinfo() to fill in a struct addrinfo structure with the needed information. getaddrinfo() takes a hints parameter, which tells it what we're looking for. In this case, we've zeroed out hints using memset() first. Then, we set ai_family = AF_INET. AF_INET specifies that we are looking for an IPv4 address. We could use AF_INET6 to make our web server listen on an IPv6 address instead (more on this later).

Next, we set `ai_socktype = SOCK_STREAM`. This indicates that we're going to be using TCP. `SOCK_DGRAM` would be used if we were doing a UDP server instead.

Finally, `ai_flags = AI_PASSIVE` is set. This is telling `getaddrinfo()` that we want it to bind to the wildcard address. That is, we are asking `getaddrinfo()` to set up the address, so we listen on any available network interface.

Once `hints` is set up properly, we declare a pointer to a `struct addrinfo` structure, which holds the return information from `getaddrinfo()`. We then call the `getaddrinfo()` function. The `getaddrinfo()` function has many uses, but for our purpose, it generates an address that's suitable for `bind()`. To make it generate this, we must pass in the first parameter as `NULL` and have the `AI_PASSIVE` flag set in `hints.ai_flags`.

The second parameter to `getaddrinfo()` is the port we listen for connections on. A standard HTTP server would use port `80`. However, only privileged users on Unix-like operating systems can bind to ports `0` through `1023`. The choice of port number here is arbitrary, but we use `8080` to avoid issues. If you are running with superuser privileges, feel free to change the port number to `80` if you like. Keep in mind that only one program can bind to a particular port at a time. If you try to use a port that is already in use, then the call to `bind()` fails. In this case, just change the port number to something else and try again.

It is common to see programs that don't use `getaddrinfo()` here. Instead, they fill in a `struct addrinfo` structure directly. The advantage to using `getaddrinfo()` is that it is protocol-independent. Using `getaddrinfo()` makes it very easy to convert our program from IPv4 to IPv6. In fact, we only need to change `AF_INET` to `AF_INET6`, and our program will work on IPv6. If we filled in the `struct addrinfo` structure directly, we would need to make many tedious changes to convert our program into IPv6.

Now that we've figured out our local address info, we can create the socket:

```
/*time_server.c continued*/

    printf("Creating socket...\n");
    SOCKET socket_listen;
    socket_listen = socket(bind_address->ai_family,
            bind_address->ai_socktype, bind_address->ai_protocol);
```

Here, we define `socket_listen` as a `SOCKET` type. Recall that `SOCKET` is a Winsock type on Windows, and that we have a macro defining it as `int` on other platforms. We call the `socket()` function to generate the actual socket. `socket()` takes three parameters: the socket family, the socket type, and the socket protocol. The reason we used `getaddrinfo()` before calling `socket()` is that we can now pass in parts of `bind_address` as the arguments to `socket()`. Again, this makes it very easy to change our program's protocol without needing a major rewrite.

It is common to see programs written so that they call `socket()` first. The problem with this is that it makes the program more complicated as the socket family, type, and protocol must be entered multiple times. Structuring our program as we have here is better.

We should check that the call to `socket()` was successful:

```
/*time_server.c continued*/

    if (!ISVALIDSOCKET(socket_listen)) {
        fprintf(stderr, "socket() failed. (%d)\n", GETSOCKETERRNO());
         return 1;
    }
```

We can check that `socket_listen` is valid using the `ISVALIDSOCKET()` macro we defined earlier. If the socket is not valid, we print an error message. Our `GETSOCKETERRNO()` macro is used to retrieve the error number in a cross-platform way.

After the socket has been created successfully, we can call `bind()` to associate it with our address from `getaddrinfo()`:

```
/*time_server.c continued*/

    printf("Binding socket to local address...\n");
    if (bind(socket_listen,
                bind_address->ai_addr, bind_address->ai_addrlen)) {
        fprintf(stderr, "bind() failed. (%d)\n", GETSOCKETERRNO());
        return 1;
    }
    freeaddrinfo(bind_address);
```

`bind()` returns 0 on success and non-zero on failure. If it fails, we print the error number much like we did for the error handling on `socket()`. `bind()` fails if the port we are binding to is already in use. In that case, either close the program using that port or change your program to use a different port.

After we have bound to `bind_address`, we can call the `freeaddrinfo()` function to release the address memory.

Once the socket has been created and bound to a local address, we can cause it to start listening for connections with the `listen()` function:

```
/*time_server.c continued*/

    printf("Listening...\n");
    if (listen(socket_listen, 10) < 0) {
        fprintf(stderr, "listen() failed. (%d)\n", GETSOCKETERRNO());
        return 1;
    }
```

The second argument to `listen()`, which is `10` in this case, tells `listen()` how many connections it is allowed to queue up. If many clients are connecting to our server all at once, and we aren't dealing with them fast enough, then the operating system begins to queue up these incoming connections. If `10` connections become queued up, then the operating system will reject new connections until we remove one from the existing queue.

Error handling for `listen()` is done the same way as we did for `bind()` and `socket()`.

After the socket has begun listening for connections, we can accept any incoming connection with the `accept()` function:

```
/*time_server.c continued*/

    printf("Waiting for connection...\n");
    struct sockaddr_storage client_address;
    socklen_t client_len = sizeof(client_address);
    SOCKET socket_client = accept(socket_listen,
            (struct sockaddr*) &client_address, &client_len);
    if (!ISVALIDSOCKET(socket_client)) {
        fprintf(stderr, "accept() failed. (%d)\n", GETSOCKETERRNO());
        return 1;
    }
```

`accept()` has a few functions. First, when it's called, it will block your program until a new connection is made. In other words, your program will sleep until a connection is made to the listening socket. When the new connection is made, `accept()` will create a new socket for it. Your original socket continues to listen for new connections, but the new socket returned by `accept()` can be used to send and receive data over the newly established connection. `accept()` also fills in address info of the client that connected.

Before calling `accept()`, we must declare a new `struct sockaddr_storage` variable to store the address info for the connecting client. The `struct sockaddr_storage` type is guaranteed to be large enough to hold the largest supported address on your system. We must also tell `accept()` the size of the address buffer we're passing in. When `accept()` returns, it will have filled in `client_address` with the connected client's address and `client_len` with the length of that address. `client_len` differs, depending on whether the connection is using IPv4 or IPv6.

We store the return value of `accept()` in `socket_client`. We check for errors by detecting if `client_socket` is a valid socket. This is done in exactly the same way as we did for `socket()`.

At this point, a TCP connection has been established to a remote client. We can print the client's address to the console:

```
/*time_server.c continued*/

    printf("Client is connected... ");
    char address_buffer[100];
    getnameinfo((struct sockaddr*)&client_address,
            client_len, address_buffer, sizeof(address_buffer), 0, 0,
            NI_NUMERICHOST);
    printf("%s\n", address_buffer);
```

This step is completely optional, but it is good practice to log network connections somewhere.

`getnameinfo()` takes the client's address and address length. The address length is needed because `getnameinfo()` can work with both IPv4 and IPv6 addresses. We then pass in an output buffer and buffer length. This is the buffer that `getnameinfo()` writes its hostname output to. The next two arguments specify a second buffer and its length. `getnameinfo()` outputs the service name to this buffer. We don't care about that, so we've passed in 0 for those two parameters. Finally, we pass in the NI_NUMERICHOST flag, which specifies that we want to see the hostname as an IP address.

As we are programming a web server, we expect the client (for example, a web browser) to send us an HTTP request. We read this request using the `recv()` function:

```
/*time_server.c continued*/

    printf("Reading request...\n");
    char request[1024];
    int bytes_received = recv(socket_client, request, 1024, 0);
    printf("Received %d bytes.\n", bytes_received);
```

We define a request buffer, so that we can store the browser's HTTP request. In this case, we allocate 1,024 bytes to it, which should be enough for this application. `recv()` is then called with the client's socket, the request buffer, and the request buffer size. `recv()` returns the number of bytes that are received. If nothing has been received yet, `recv()` blocks until it has something. If the connection is terminated by the client, `recv()` returns 0 or -1, depending on the circumstance. We are ignoring that case here for simplicity, but you should always check that `recv() > 0` in production. The last parameter to `recv()` is for flags. Since we are not doing anything special, we simply pass in 0.

The request received from our client should follow the proper HTTP protocol. We will go into detail about HTTP in Chapter 6, *Building a Simple Web Client*, and Chapter 7, *Building a Simple Web Server*, where we will work on web clients and servers. A real web server would need to parse the request and look at which resource the web browser is requesting. Our web server only has one function—to tell us what time it is. So, for now, we just ignore the request altogether.

If you want to print the browser's request to the console, you can do it like this:

```
printf("%.*s", bytes_received, request);
```

Note that we use the `printf()` format string, `"%.*s"`. This tells `printf()` that we want to print a specific number of characters—bytes_received. It is a common mistake to try printing data that's received from `recv()` directly as a C string. There is no guarantee that the data received from `recv()` is null terminated! If you try to print it with `printf(request)` or `printf("%s", request)`, you will likely receive a segmentation fault error (or at best it will print some garbage).

Now that the web browser has sent its request, we can send our response back:

```
/*time_server.c continued*/

    printf("Sending response...\n");
    const char *response =
        "HTTP/1.1 200 OK\r\n"
        "Connection: close\r\n"
        "Content-Type: text/plain\r\n\r\n"
        "Local time is: ";
    int bytes_sent = send(socket_client, response, strlen(response), 0);
    printf("Sent %d of %d bytes.\n", bytes_sent, (int)strlen(response));
```

To begin with, we set `char *response` to a standard HTTP response header and the beginning of our message (`Local time is:`). We will discuss HTTP in detail in `Chapter 6`, *Building a Simple Web Client*, and `Chapter 7`, *Building a Simple Web Server*. For now, know that this response tells the browser three things—your request is OK; the server will close the connection when all the data is sent and the data you receive will be plain text.

The HTTP response header ends with a blank line. HTTP requires line endings to take the form of a carriage return character, followed by a newline character. So, a blank line in our response is `\r\n`. The part of the string that comes after the blank line, `Local time is:`, is treated by the browsers as plain text.

We send the data to the client using the `send()` function. This function takes the client's socket, a pointer to the data to be sent, and the length of the data to send. The last parameter to `send()` is flags. We don't need to do anything special, so we pass in `0`.

`send()` returns the number of bytes sent. You should generally check that the number of bytes sent was as expected, and you should attempt to send the rest if it's not. We are ignoring that detail here for simplicity. (Also, we are only attempting to send a few bytes; if `send()` can't handle that, then something is probably very broken, and resending won't help.)

After the HTTP header and the beginning of our message is sent, we can send the actual time. We get the local time the same way we did in `time_console.c`, and we send it using `send()`:

```
/*time_server.c continued*/

    time_t timer;
    time(&timer);
    char *time_msg = ctime(&timer);
    bytes_sent = send(socket_client, time_msg, strlen(time_msg), 0);
    printf("Sent %d of %d bytes.\n", bytes_sent, (int)strlen(time_msg));
```

We must then close the client connection to indicate to the browser that we've sent all of our data:

```
/*time_server.c continued*/

    printf("Closing connection...\n");
    CLOSESOCKET(socket_client);
```

If we don't close the connection, the browser will just wait for more data until it times out.

At this point, we could call `accept()` on `socket_listen` to accept additional connections. That is exactly what a real server would do. However, as this is just a quick example program, we will instead close the listening socket too and terminate the program:

```
/*time_server.c continued*/

    printf("Closing listening socket...\n");
    CLOSESOCKET(socket_listen);


#if defined(_WIN32)
    WSACleanup();
#endif


    printf("Finished.\n");

    return 0;
}
```

That's the complete program. After you compile and run it, you can navigate a web browser to it, and it'll display the current time.

On Linux and macOS, you can compile and run the program like this:

```
gcc time_server.c -o time_server
./time_server
```

On Windows, you can compile and run with MinGW using these commands:

```
gcc time_server.c -o time_server.exe -lws2_32
time_server
```

When you run the program, it waits for a connection. You can open a web browser and navigate to `http://127.0.0.1:8080` to load the web page. Recall that `127.0.0.1` is the IPv4 loopback address, which connects to the same machine it's running on. The `:8080` part of the URL specifies the port number to connect to. If it were left out, your browser would default to port `80`, which is the standard for HTTP connections.

Here is what you should see if you compile and run the program, and then connect a web browser to it on the same computer:

```
m1:Desktop honp$ gcc time_server.c -o time_server
m1:Desktop honp$ ./time_server
Configuring local address...
Creating socket...
Binding socket to local address...
Listening...
Waiting for connection...
Client is connected... 127.0.0.1
Reading request...
Received 320 bytes.
Sending response...
Sent 79 of 79 bytes.
Sent 25 of 25 bytes.
Closing connection...
Closing listening socket...
Finished.
m1:Desktop honp$
```

Here is the web browser connected to our `time_server` program on port `8080`:

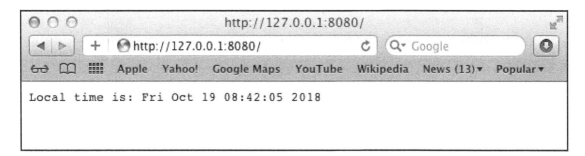

```
Local time is: Fri Oct 19 08:42:05 2018
```

Working with IPv6

Please recall the `hints.ai_family = AF_INET` part of `time_server.c` near the beginning of the `main()` function. If this line is changed to `hints.ai_family = AF_INET6`, then your web server listens for IPv6 connections instead of IPv4 connections. This modified file is included in the GitHub repository as `time_server_ipv6.c`.

In this case, you should navigate your web browser to `http://[::1]:8080` to see the web page. `::1` is the IPv6 loopback address, which tells the web browser to connect to the same machine it's running on. In order to use IPv6 addresses in URLs, you need to put them in square brackets, `[]`. `:8080` specifies the port number in the same way that we did for the IPv4 example.

Here is what you should see when compiling, running, and connecting a web browser to our `time_server_ipv6` program:

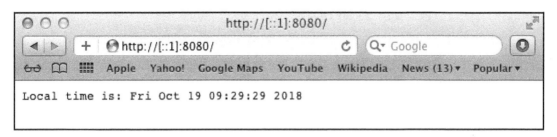

```
m1:Desktop honp$ gcc time_server.c -o time_server
m1:Desktop honp$ ./time_server
Configuring local address...
Creating socket...
Binding socket to local address...
Listening...
Waiting for connection...
Client is connected... ::1
Reading request...
Received 316 bytes.
Sending response...
Sent 79 of 79 bytes.
Sent 25 of 25 bytes.
Closing connection...
Closing listening socket...
Finished.
m1:Desktop honp$
```

Here is the web browser that's connected to our server using an IPv6 socket:

```
Local time is: Fri Oct 19 09:29:29 2018
```

See `time_server_ipv6.c` for the complete program.

Supporting both IPv4 and IPv6

It is also possible for the listening IPv6 socket to accept IPv4 connections with a dual-stack socket. Not all operating systems support dual-stack sockets. With Linux in particular, support varies between distros. If your operating system does support dual-stack sockets, then I highly recommend implementing your server programs using this feature. It allows your programs to communicate with both IPv4 and IPv6 peers while requiring no extra work on your part.

We can modify `time_server_ipv6.c` to use dual-stack sockets with only a minor addition. After the call to `socket()` and before the call to `bind()`, we must clear the `IPV6_V6ONLY` flag on the socket. This is done with the `setsockopt()` function:

```
/*time_server_dual.c excerpt*/

    int option = 0;
    if (setsockopt(socket_listen, IPPROTO_IPV6, IPV6_V6ONLY,
(void*)&option, sizeof(option))) {
        fprintf(stderr, "setsockopt() failed. (%d)\n", GETSOCKETERRNO());
        return 1;
    }
```

We first declare `option` as an integer and set it to 0. `IPV6_V6ONLY` is enabled by default, so we clear it by setting it to 0. `setsockopt()` is called on the listening socket. We pass in `IPPROTO_IPV6` to tell it what part of the socket we're operating on, and we pass in `IPV6_V6ONLY` to tell it which flag we are setting. We then pass in a pointer to our option and its length. `setsockopt()` returns 0 on success.

Windows Vista and later supports dual-stack sockets. However, many Windows headers are missing the definitions for `IPV6_V6ONLY`. For this reason, it might make sense to include the following code snippet at the top of the file:

```
/*time_server_dual.c excerpt*/

#if !defined(IPV6_V6ONLY)
#define IPV6_V6ONLY 27
#endif
```

Keep in mind that the socket needs to be initially created as an IPv6 socket. This is accomplished with the `hints.ai_family = AF_INET6` line in our code.

When an IPv4 peer connects to our dual-stack server, the connection is remapped to an IPv6 connection. This happens automatically and is taken care of by the operating system. When your program sees the client IP address, it will still be presented as a special IPv6 address. These are represented by IPv6 addresses where the first 96 bits consist of the prefix—0:0:0:0:0:ffff. The last 32 bits of the address are used to store the IPv4 address. For example, if a client connects with the IPv4 address 192.168.2.107, then your dual-stack server sees it as the IPv6 address ::ffff.192.168.2.107.

Here is what it looks like to compile, run, and connect to time_server_dual:

```
● ○ ○                    🖥 Desktop — bash — 80×20
m1:Desktop honp$ gcc time_server_dual.c -o time_server_dual
m1:Desktop honp$ ./time_server_dual
Configuring local address...
Creating socket...
Binding socket to local address...
Listening...
Waiting for connection...
Client is connected... ::ffff:127.0.0.1
Reading request...
Received 320 bytes.
Sending response...
Sent 79 of 79 bytes.
Sent 25 of 25 bytes.
Closing connection...
Closing listening socket...
Finished.
m1:Desktop honp$ ›
```

Here is a web browser connected to our time_server_dual program using the loopback IPv4 address:

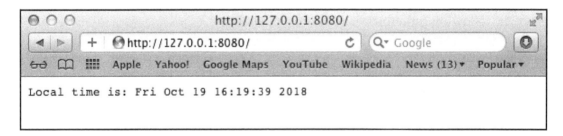

Notice that the browser is navigating to the IPv4 address 127.0.0.1, but we can see on the console that the server sees the connection as coming from the IPv6 address ::ffff:127.0.0.1.

See time_server_dual.c for the complete dual-stack socket server.

Networking with inetd

On Unix-like systems, such as Linux or macOS, a service called *inetd* can be used to turn console-only applications into networked ones. You can configure *inetd* (with `/etc/inetd.conf`) with your program's location, port number, protocol (TCP or UDP), and the user you want it to run as. *inetd* will then listen for connections on your desired port. After an incoming connection is accepted by *inetd*, it will start your program and redirect all socket input/output through `stdin` and `stdout`.

Using *inetd*, we could have `time_console.c` behave like `time_server.c` with very minimal changes. We would only need to add in an extra `printf()` function with the HTTP response header, read from `stdin`, and configure *inetd*.

You may be able to use *inetd* on Windows through Cygwin or the Windows Subsystem for Linux.

Summary

In this chapter, we learned about the basics of using sockets for network programming. Although there are many differences between Berkeley sockets (used on Unix-like operating systems) and Winsock sockets (used on Windows), we mitigated those differences with preprocessor statements. In this way, it was possible to write one program that compiles cleanly on Windows, Linux, and macOS.

We covered how the UDP protocol is connectionless and what that means. We learned that TCP, being a connection-oriented protocol, gives some reliability guarantees, such as automatically detecting and resending lost packets. We also saw that UDP is often used for simple protocols (for example, DNS) and for real-time streaming applications. TCP is used for most other protocols.

After that, we worked through a real example by converting a console application into a web server. We learned how to write the program using the `getaddrinfo()` function, and why that matters for making the program IPv4/IPv6-agnostic. We used `bind()`, `listen()`, and `accept()` on the server to wait for an incoming connection from the web browser. Data was then read from the client using `recv()`, and a reply was sent using `send()`. Finally, we terminated the connection with `close()` (`closesocket()` on Windows).

When we built the web server, `time_server.c`, we covered much ground. It's OK if you didn't understand all of it. We will revisit many of these functions again throughout `Chapter 3`, *An In-Depth Overview of TCP Connections*, and the rest of this book.

In the next chapter, `Chapter 3`, *An In-Depth Overview of TCP Connections*, we will consider programming for TCP connections in more depth.

Questions

Try these questions to test your knowledge on this chapter:

1. What is a socket?
2. What is a connectionless protocol? What is a connection-oriented protocol?
3. Is UDP a connectionless or connection-oriented protocol?
4. Is TCP a connectionless or connection-oriented protocol?
5. What types of applications generally benefit from using the UDP protocol?
6. What types of applications generally benefit from using the TCP protocol?
7. Does TCP guarantee that data will be transmitted successfully?
8. What are some of the main differences between Berkeley sockets and Winsock sockets?
9. What does the `bind()` function do?
10. What does the `accept()` function do?
11. In a TCP connection, does the client or the server send application data first?

Answers are in `Appendix A`, *Answers to Questions*.

An In-Depth Overview of TCP Connections

3

In `Chapter 2`, *Getting to Grips with Socket APIs*, we implemented a simple TCP server that served a web page with HTTP. In this chapter, we will begin by implementing a TCP client. This client is able to establish an IPv4 or IPv6 TCP connection with any listening TCP server. It will be a useful debugging tool that we can reuse in the rest of this book.

Our TCP server from the last chapter was limited to accepting only one connection. In this chapter, we will look at multiplexing techniques to allow our programs to handle many separate connections simultaneously.

The following topics are covered in this chapter:

- Configuring a remote address with `getaddrinfo()`
- Initiating a TCP connection with `connect()`
- Detecting terminal input in a non-blocking manner
- Multiplexing with `fork()`
- Multiplexing with `select()`
- Detecting peer disconnects
- Implementing a very basic microservice
- The stream-like nature of TCP
- The blocking behavior of `send()`

Technical requirements

The example programs for this chapter can be compiled with any modern C compiler. We recommend MinGW on Windows and GCC on Linux and macOS. See Appendix B, *Setting Up Your C Compiler On Windows*, Appendix C, *Setting Up Your C Compiler On Linux*, and Appendix D, *Setting Up Your C Compiler On macOS*, for compiler setup.

The code for this book can be found in this book's GitHub repository: https://github.com/codeplea/Hands-On-Network-Programming-with-C.

From the command line, you can download the code for this chapter with the following command:

```
git clone https://github.com/codeplea/Hands-On-Network-Programming-with-C
cd Hands-On-Network-Programming-with-C/chap03
```

Each example program in this chapter runs on Windows, Linux, and macOS. While compiling on Windows, each example program requires that you link with the Winsock library. This can be accomplished by passing the -lws2_32 option to gcc.

We provide the exact commands that are needed to compile each example as it is introduced.

All of the example programs in this chapter require the same header files and C macros that we developed in Chapter 2, *Getting to Grips with Socket APIs*. For brevity, we put these statements in a separate header file, chap03.h, which we can include in each program. For an explanation of these statements, please refer to Chapter 2, *Getting to Grips with Socket APIs*.

The contents of chap03.h is as follows:

```
/*chap03.h*/

#if defined(_WIN32)
#ifndef _WIN32_WINNT
#define _WIN32_WINNT 0x0600
#endif
#include <winsock2.h>
#include <ws2tcpip.h>
#pragma comment(lib, "ws2_32.lib")

#else
#include <sys/types.h>
#include <sys/socket.h>
#include <netinet/in.h>
```

```
#include <arpa/inet.h>
#include <netdb.h>
#include <unistd.h>
#include <errno.h>

#endif


#if defined(_WIN32)
#define ISVALIDSOCKET(s) ((s) != INVALID_SOCKET)
#define CLOSESOCKET(s) closesocket(s)
#define GETSOCKETERRNO() (WSAGetLastError())

#else
#define ISVALIDSOCKET(s) ((s) >= 0)
#define CLOSESOCKET(s) close(s)
#define SOCKET int
#define GETSOCKETERRNO() (errno)
#endif


#include <stdio.h>
#include <string.h>
```

Multiplexing TCP connections

The socket APIs are blocking by default. When you use `accept()` to wait for an incoming connection, your program's execution is blocked until a new incoming connection is actually available. When you use `recv()` to read incoming data, your program's execution blocks until new data is actually available.

In the last chapter, we built a simple TCP server. This server only accepted one connection, and it only read data from that connection once. Blocking wasn't a problem then, because our server had no other purpose than to serve its one and only client.

In the general case, though, blocking I/O can be a significant problem. Imagine that our server from Chapter 2, *Getting to Grips with Socket APIs*, needed to serve multiple clients. Then, imagine that one slow client connected to it. Maybe this slow client takes a minute before sending its first data. During this minute, our server would simply be waiting on the `recv()` call to return. If other clients were trying to connect, they would have to wait it out.

Blocking on `recv()` like this isn't really acceptable. A real application usually needs to be able to manage several connections simultaneously. This is obviously true on the server side, as most servers are built to manage many connected clients. Imagine running a website where hundreds of clients are connected at once. Serving these clients one at a time would be a non-starter.

Blocking also isn't usually acceptable on the client side either. If you imagine building a fast web browser, it needs to be able to download many images, scripts, and other resources in parallel. Modern web browsers also have a **tab** feature where many whole web pages can be loaded in parallel.

What we need is a technique for handling many separate connections simultaneously.

Polling non-blocking sockets

It is possible to configure sockets to use a non-blocking operation. One way to do this is by calling `fcntl()` with the `O_NONBLOCK` flag (`ioctlsocket()` with the `FIONBIO` flag on Windows), although other ways also exist. Once in non-blocking mode, a call to `recv()` with no data will return immediately. See `Chapter 13`, *Socket Programming Tips and Pitfalls*, for more information.

A program structured with this in mind could simply check each of its active sockets in turn, continuously. It would handle any socket that returned data and ignore any socket that didn't. This is called **polling**. Polling can be a waste of computer resources since most of the time, there will be no data to read. It also complicates the program somewhat, as the programmer is required to manually track which sockets are active and which state, they are in. Return values from `recv()` must also be handled differently than with blocking sockets.

For these reasons, we won't use polling in this book.

Forking and multithreading

Another possible solution to multiplexing socket connections is to start a new thread or process for each connection. In this case, blocking sockets are fine, as they block only their servicing thread/process, and they do not block other threads/processes. This can be a useful technique, but it also has some downsides. First of all, threading is tricky to get right. This is especially true if the connections must share any state between them. It is also less portable as each operating system provides a different API for these features.

On Unix-based systems, such as Linux and macOS, starting a new process is very easy. We simply use the `fork()` function. The `fork()` function splits the executing program into two separate processes. A multi-process TCP server may accept connections like this:

```
while(1) {
    socket_client = accept(socket_listen, &new_client, &new_client_length);
    int pid = fork();
    if (pid == 0) { //child process
        close(socket_listen);
        recv(socket_client, ...);
        send(socket_client, ...);
        close(socket_client);
        exit(0);
    }
    //parent process
    close(socket_client);
}
```

In this example, the program blocks on `accept()`. When a new connection is established, the program calls `fork()` to split into two processes. The child process, where `pid == 0`, only services this one connection. Therefore, the child process can use `recv()` freely without worrying about blocking. The parent process simply calls `close()` on the new connection and returns to listening for more connections with `accept()`.

Using multiple processes/threads is much more complicated on Windows. Windows provides `CreateProcess()`, `CreateThread()`, and many other functions for these features. However—and I can say this objectively—they are all much harder to use than Unix's `fork()`.

Debugging these multi-process/thread programs can be much more difficult compared to the single process case. Communicating between sockets and managing shared state is also much more burdensome. For these reasons, we will avoid `fork()` and other multi-process/thread techniques for the rest of this book.

That being said, an example TCP server using fork is included in this chapter's code. It's named `tcp_serve_toupper_fork.c`. It does not run on Windows, but it should compile and run cleanly on Linux and macOS. I would suggest finishing the rest of this chapter before looking at it.

The select() function

Our preferred technique for multiplexing is to use the `select()` function. We can give `select()` a set of sockets, and it tells us which ones are ready to be read. It can also tell us which sockets are ready to write to and which sockets have exceptions. Furthermore, it is supported by both Berkeley sockets and Winsock. Using `select()` keeps our programs portable.

Synchronous multiplexing with select()

The `select()` function has several useful features. Given a set of sockets, it can be used to block until any of the sockets in that set is ready to be read from. It can also be configured to return if a socket is ready to be written to or if a socket has an error. Additionally, we can configure `select()` to return after a specified time if none of these events take place.

The C function prototype for `select()` is as follows:

```
int select(int nfds, fd_set *readfds, fd_set *writefds,
    fd_set *exceptfds, struct timeval *timeout);
```

Before calling `select()`, we must first add our sockets into an `fd_set`. If we have three sockets, `socket_listen`, `socket_a`, and `socket_b`, we add them to an `fd_set`, like this:

```
fd_set our_sockets;
FD_ZERO(&our_sockets);
FD_SET(socket_listen, &our_sockets);
FD_SET(socket_a, &our_sockets);
FD_SET(socket_b, &our_sockets);
```

It is important to zero-out the `fd_set` using `FD_ZERO()` before use.

Socket descriptors are then added to the `fd_set` one at a time using `FD_SET()`. A socket can be removed from an `fd_set` using `FD_CLR()`, and we can check for the presence of a socket in the set using `FD_ISSET()`.

You may see some programs manipulating an `fd_set` directly. I recommend that you use only `FD_ZERO()`, `FD_SET()`, `FD_CLR()`, and `FD_ISSET()` to maintain portability between Berkeley sockets and Winsock.

select() also requires that we pass a number that's larger than the largest socket descriptor we are going to monitor. (This parameter is ignored on Windows, but we will always do it anyway for portability.) We store the largest socket descriptor in a variable, like this:

```
SOCKET max_socket;
max_socket = socket_listen;
if (socket_a > max_socket) max_socket = socket_a;
if (socket_b > max_socket) max_socket = socket_b;
```

When we call select(), it modifies our fd_set of sockets to indicate which sockets are ready. For that reason, we want to copy our socket set before calling it. We can copy an fd_set with a simple assignment like this, and then call select() like this:

```
fd_set copy;
copy = our_sockets;

select(max_socket+1, &copy, 0, 0, 0);
```

This call blocks until at least one of the sockets is ready to be read from. When select() returns, copy is modified so that it only contains the sockets that are ready to be read from. We can check which sockets are still in copy using FD_ISSET(), like this:

```
if (FD_ISSET(socket_listen, &copy)) {
    //socket_listen has a new connection
    accept(socket_listen...
}

if (FD_ISSET(socket_a, &copy)) {
    //socket_a is ready to be read from
    recv(socket_a...
}

if (FD_ISSET(socket_b, &copy)) {
    //socket_b is ready to be read from
    recv(socket_b...
}
```

In the previous example, we passed our fd_set as the second argument to select(). If we wanted to monitor an fd_set for writability instead of readability, we would pass our fd_set as the third argument to select(). Likewise, we can monitor a set of sockets for exceptions by passing it as the fourth argument to select().

select() timeout

The last argument taken by `select()` allows us to specify a timeout. It expects a pointer to `struct timeval`. The `timeval` structure is declared as follows:

```
struct timeval {
    long tv_sec;
    long tv_usec;
}
```

`tv_sec` holds the number of seconds, and `tv_usec` holds the number of microseconds (1,000,000th second). If we want `select()` to wait a maximum of 1.5 seconds, we can call it like this:

```
struct timeval timeout;
timeout.tv_sec = 1;
timeout.tv_usec = 500000;
select(max_socket+1, &copy, 0, 0, &timeout);
```

In this case, `select()` returns after a socket in `fd_set copy` is ready to read or after 1.5 seconds has elapsed, whichever is sooner.

If `timeout.tv_sec = 0` and `timeout.tv_usec = 0`, then `select()` returns immediately (after changing the `fd_set` as appropriate). As we saw previously, if we pass in a null pointer for the timeout parameter, then `select()` does not return until at least one socket is ready to be read.

`select()` can also be used to monitor for writeable sockets (sockets where we could call `send()` without blocking), and sockets with exceptions. We can check for all three conditions with one call:

```
select(max_sockets+1, &ready_to_read, &ready_to_write, &excepted,
    &timeout);
```

On success, `select()` itself returns the number of socket descriptors contained in the (up to) three descriptor sets it monitored. The return value is zero if it timed out before any sockets were readable/writeable/excepted. `select()` returns –1 to indicate an error.

Iterating through an fd_set

We can iterate through an fd_set using a simple for loop. Essentially, we start at 1, since all socket descriptors are positive numbers, and we continue through to the largest known socket descriptor in the set. For each possible socket descriptor, we simply use FD_ISSET() to check if it is in the set. If we wanted to call CLOSESOCKET() for every socket in the fd_set master, we could do it like this:

```
SOCKET i;
for (i = 1; i <= max_socket; ++i) {
    if (FD_ISSSET(i, &master)) {
        CLOSESOCKET(i);
    }
}
```

This may seem like a brute-force approach, and it actually kind of is. However, these are the tools that we have to work with. FD_ISSET() runs very fast, and it's likely that processor time spent on other socket operations will dwarf what time was spent iterating through them in this manner. Nevertheless, you may be able to optimize this operation by additionally storing your sockets in an array or linked list. I don't recommend that you make this optimization unless you profile your code and find the simple for loop iteration to be a significant bottleneck.

select() on non-sockets

On Unix-based systems, select() can also be used on file and terminal I/O, which can be extremely useful. This doesn't work on Windows, though. Windows only supports select() for sockets.

A TCP client

It will be useful for us to have a TCP client that can connect to any TCP server. This TCP client will take in a hostname (or IP address) and port number from the command line. It will attempt a connection to the TCP server at that address. If successful, it will relay data that's received from that server to the terminal and data inputted into the terminal to the server. It will continue until either it is terminated (with *Ctrl + C*) or the server closes the connection.

This is useful as a learning opportunity to see how to program a TCP client, but it is also useful for testing the TCP server programs we develop throughout this book.

Our basic program flow looks like this:

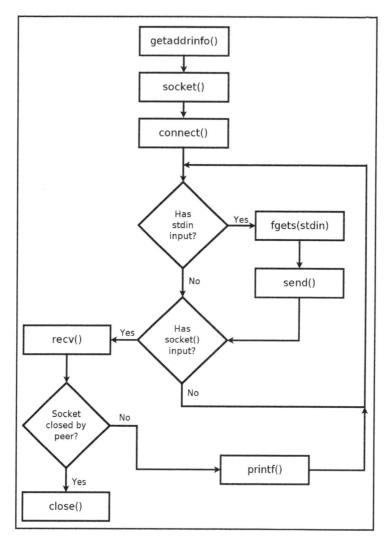

Our program first uses getaddrinfo() to resolve the server address from the command-line arguments. Then, the socket is created with a call to socket(). The fresh socket has connect() called on it to connect to the server. We use select() to monitor for socket input. select() also monitors for terminal/keyboard input on non-Windows systems. On Windows, we use the _kbhit() function to detect terminal input. If terminal input is available, we send it over the socket using send(). If select() indicated that socket data is available, we read it with recv() and display it to the terminal. This select() loop is repeated until the socket is closed.

TCP client code

We begin our TCP client by including the header file, chap03.h, which was printed at the beginning of this chapter. This header file includes the various other headers and macros we need for cross-platform networking:

```
/*tcp_client.c*/

#include "chap03.h"
```

On Windows, we also need the conio.h header. This is required for the _kbhit() function, which helps us by indicating whether terminal input is waiting. We conditionally include this header, like so:

```
/*tcp_client.c*/

#if defined(_WIN32)
#include <conio.h>
#endif
```

We can then begin the main() function and initialize Winsock:

```
/*tcp_client.c*/

int main(int argc, char *argv[]) {

#if defined(_WIN32)
    WSADATA d;
    if (WSAStartup(MAKEWORD(2, 2), &d)) {
        fprintf(stderr, "Failed to initialize.\n");
        return 1;
    }
#endif
```

We would like our program to take the hostname and port number of the server it should connect to as command-line arguments. This makes our program flexible. We have our program check that these command-line arguments are given. If they aren't, it displays usage information:

```
/*tcp_client.c*/

    if (argc < 3) {
        fprintf(stderr, "usage: tcp_client hostname port\n");
        return 1;
    }
```

argc contains the number of argument values available to us. Because the first argument is always our program's name, we check that there is a total of at least three arguments. The actual values themselves are stored in argv[].

We then use these values to configure a remote address for connection:

```
/*tcp_client.c*/

    printf("Configuring remote address...\n");
    struct addrinfo hints;
    memset(&hints, 0, sizeof(hints));
    hints.ai_socktype = SOCK_STREAM;
    struct addrinfo *peer_address;
    if (getaddrinfo(argv[1], argv[2], &hints, &peer_address)) {
        fprintf(stderr, "getaddrinfo() failed. (%d)\n", GETSOCKETERRNO());
        return 1;
    }
```

This is similar to how we called getaddrinfo() in Chapter 2, *Getting to Grips with Socket APIs*. However, in Chapter 2, *Getting to Grips with Socket APIs*, we wanted it to configure a local address, whereas this time, we want it to configure a remote address.

We set hints.ai_socktype = SOCK_STREAM to tell getaddrinfo() that we want a TCP connection. Remember that we could set SOCK_DGRAM to indicate a UDP connection.

In Chapter 2, *Getting to Grips with Socket APIs*, we also set the family. We don't need to set the family here, as we can let getaddrinfo() decide if IPv4 or IPv6 is the proper protocol to use.

For the call to getaddrinfo() itself, we pass in the hostname and port as the first two arguments. These are passed directly in from the command line. If they aren't suitable, then getaddrinfo() returns non-zero and we print an error message. If everything goes well, then our remote address is in the peer_address variable.

`getaddrinfo()` is very flexible about how it takes inputs. The hostname could be a domain name like `example.com` or an IP address such as `192.168.17.23` or `::1`. The port can be a number, such as `80`, or a protocol, such as `http`.

After `getaddrinfo()` configures the remote address, we print it out. This isn't really necessary, but it is a good debugging measure. We use `getnameinfo()` to convert the address back into a string, like this:

```
/*tcp_client.c*/

    printf("Remote address is: ");
    char address_buffer[100];
    char service_buffer[100];
    getnameinfo(peer_address->ai_addr, peer_address->ai_addrlen,
            address_buffer, sizeof(address_buffer),
            service_buffer, sizeof(service_buffer),
            NI_NUMERICHOST);
    printf("%s %s\n", address_buffer, service_buffer);
```

We can then create our socket:

```
/*tcp_client.c*/

    printf("Creating socket...\n");
    SOCKET socket_peer;
    socket_peer = socket(peer_address->ai_family,
            peer_address->ai_socktype, peer_address->ai_protocol);
    if (!ISVALIDSOCKET(socket_peer)) {
        fprintf(stderr, "socket() failed. (%d)\n", GETSOCKETERRNO());
        return 1;
    }
```

This call to `socket()` is done in exactly the same way as it was in Chapter 2, *Getting to Grips with Socket APIs*. We use `peer_address` to set the proper socket family and protocols. This keeps our program very flexible, as the `socket()` call creates an IPv4 or IPv6 socket as needed.

After the socket has been created, we call `connect()` to establish a connection to the remote server:

```
/*tcp_client.c */

    printf("Connecting...\n");
    if (connect(socket_peer,
                peer_address->ai_addr, peer_address->ai_addrlen)) {
        fprintf(stderr, "connect() failed. (%d)\n", GETSOCKETERRNO());
```

```
        return 1;
    }
    freeaddrinfo(peer_address);
```

`connect()` takes three arguments—the socket, the remote address, and the remote address length. It returns 0 on success, so we print an error message if it returns non-zero. This call to `connect()` is extremely similar to how we called `bind()` in Chapter 2, *Getting to Grips with Socket APIs*. Where `bind()` associates a socket with a local address, `connect()` associates a socket with a remote address and initiates the TCP connection.

After we've called `connect()` with `peer_address`, we use the `freeaddrinfo()` function to free the memory for `peer_address`.

If we've made it this far, then a TCP connection has been established to the remote server. We let the user know by printing a message and instructions on how to send data:

```
/*tcp_client.c */

    printf("Connected.\n");
    printf("To send data, enter text followed by enter.\n");
```

Our program should now loop while checking both the terminal and socket for new data. If new data comes from the terminal, we send it over the socket. If new data is read from the socket, we print it out to the terminal.

It is clear we cannot call `recv()` directly here. If we did, it would block until data comes from the socket. In the meantime, if our user enters data on the terminal, that input is ignored. Instead, we use `select()`. We begin our loop and set up the call to `select()`, like this:

```
/*tcp_client.c */

    while(1) {

        fd_set reads;
        FD_ZERO(&reads);
        FD_SET(socket_peer, &reads);
#if !defined(_WIN32)
        FD_SET(0, &reads);
#endif

        struct timeval timeout;
        timeout.tv_sec = 0;
        timeout.tv_usec = 100000;

        if (select(socket_peer+1, &reads, 0, 0, &timeout) < 0) {
```

```
fprintf(stderr, "select() failed. (%d)\n", GETSOCKETERRNO());
return 1;
}
```

First, we declare a variable, `fd_set reads`, to store our socket set. We then zero it with `FD_ZERO()` and add our only socket, `socket_peer`.

On non-Windows systems, we also use `select()` to monitor for terminal input. We add `stdin` to the `reads` set with `FD_SET(0, &reads)`. This works because 0 is the file descriptor for `stdin`. Alternatively, we could have used `FD_SET(fileno(stdin), &reads)` to the same effect.

The Windows `select()` function only works on sockets. Therefore, we cannot use `select()` to monitor for console input. For this reason, we set up a timeout to the `select()` call for 100 milliseconds (100,000 microseconds). If there is no socket activity after 100 milliseconds, `select()` returns, and we can check for terminal input manually.

After `select()` returns, we check to see whether our socket is set in `reads`. If it is, then we know to call `recv()` to read the new data. The new data is printed to the console with `printf()`:

```
/*tcp_client.c*/

        if (FD_ISSET(socket_peer, &reads)) {
            char read[4096];
            int bytes_received = recv(socket_peer, read, 4096, 0);
            if (bytes_received < 1) {
                printf("Connection closed by peer.\n");
                break;
            }
            printf("Received (%d bytes): %.*s",
                    bytes_received, bytes_received, read);
        }
```

Remember, the data from `recv()` is not null terminated. For this reason, we use the `%.*s printf()` format specifier, which prints a string of a specified length.

`recv()` normally returns the number of bytes read. If it returns less than 1, then the connection has ended, and we break out of the loop to shut it down.

After checking for new TCP data, we also need to check for terminal input:

```
/*tcp_client.c */

#if defined(_WIN32)
        if(_kbhit()) {
```

```
#else
        if(FD_ISSET(0, &reads)) {
#endif
            char read[4096];
            if (!fgets(read, 4096, stdin)) break;
            printf("Sending: %s", read);
            int bytes_sent = send(socket_peer, read, strlen(read), 0);
            printf("Sent %d bytes.\n", bytes_sent);
        }
```

On Windows, we use the _kbhit() function to indicate whether any console input is waiting. _kbhit() returns non-zero if an unhandled key press event is queued up. For Unix-based systems, we simply check if select() sets the stdin file descriptor, 0. If input is ready, we call fgets() to read the next line of input. This input is then sent over our connected socket with send().

Note that fgets() includes the newline character from the input. Therefore, our sent input always ends with a newline.

If the socket has closed, send() returns −1. We ignore this case here. This is because a closed socket causes select() to return immediately, and we notice the closed socket on the next call to recv(). This is a common paradigm in TCP socket programming to ignore errors on send() while detecting and handling them on recv(). It allows us to simplify our program by keeping our connection closing logic all in one place. Later in this chapter, we will discuss other concerns regarding send().

This select() based terminal monitoring works very well on Unix-based systems. It also works equally well if input is piped in. For example, you could use our TCP client program to send a text file with a command such as cat my_file.txt | tcp_client 192.168.54.122 8080.

The Windows terminal handling leaves a bit to be desired. Windows does not provide an easy way to tell whether stdin has input available without blocking, so we use _kbhit() as a poor proxy. However, if the user presses a non-printable key, such as an arrow key, it still triggers _kbhit(), even though there is no character to read. Also, after the first key press, our program will block on fgets() until the user presses the *Enter* key. (This doesn't happen on shells that buffer entire lines, which is common outside of Windows.) This blocking behavior is acceptable, but you should know that any received TCP data will not display until after that point. _kbhit() does not work for piped input. Doing proper piped and console input on Windows is possible, of course, but it's very complicated.

We would need to use separate functions for each (`PeekNamedPipe()` and `PeekConsoleInput()`), and the logic for handling it would be as long as this entire program! Since handling terminal input isn't the purpose of this book, we're going to accept `_kbhit()` function's limitations and move on.

At this point, our program is essentially done. We can end the `while` loop, close our socket, and clean up Winsock:

```
/*tcp_client.c */

    }

    printf("Closing socket...\n");
    CLOSESOCKET(socket_peer);

#if defined(_WIN32)
    WSACleanup();
#endif

    printf("Finished.\n");
    return 0;
}
```

That's the complete program. You can compile it on Linux and macOS like this:

```
gcc tcp_client.c -o tcp_client
```

Compiling on Windows with MinGW is done like this:

```
gcc tcp_client.c -o tcp_client.exe -lws2_32
```

To run the program, remember to pass in the remote hostname/address and port number, for example:

```
tcp_client example.com 80
```

Alternatively, you can use the following command:

```
tcp_client 127.0.0.1 8080
```

A fun way to test out the TCP client would be to connect to a live web server and send an HTTP request. For example, you could connect to `example.com` on port `80` and send the following HTTP request:

```
GET / HTTP/1.1
Host: example.com
```

You must then send a blank line to indicate the end of the request. You'll receive an HTTP response back. It might look something like this:

```
root@ubby16: /home/lv/chap03
root@ubby16:/home/lv/chap03# ./tcp_client example.com http
Configuring remote address...
Remote address is: 93.184.216.34 http
Creating socket...
Connecting...
Connected.
To send data, enter text followed by enter.
GET / HTTP/1.1
Sending: GET / HTTP/1.1
Sent 15 bytes.
Host: example.com
Sending: Host: example.com
Sent 18 bytes.

Sending:
Sent 1 bytes.
Received (1592 bytes): HTTP/1.1 200 OK
Cache-Control: max-age=604800
Content-Type: text/html; charset=UTF-8
Date: Tue, 30 Oct 2018 19:59:46 GMT
Etag: "1541025663+ident"
Expires: Tue, 06 Nov 2018 19:59:46 GMT
Last-Modified: Fri, 09 Aug 2013 23:54:35 GMT
Server: ECS (ord/4CD5)
Vary: Accept-Encoding
X-Cache: HIT
Content-Length: 1270

<!doctype html>
<html>
<head>
    <title>Example Domain</title>

    <meta charset="utf-8" />
    <meta http-equiv="Content-type" content="text/html; charset=utf-8" />
    <meta name="viewport" content="width=device-width, initial-scale=1" />
    <style type="text/css">
    body {
        background-color: #f0f0f2;
        margin: 0;
        padding: 0;
        font-family: "Open Sans", "Helvetica Neue", Helvetica, Arial, sans
-serif;

    }
```

A TCP server

Microservices have become increasingly popular in recent years. The idea of microservices is that large programming problems can be split up into many small subsystems that communicate over a network. For example, if your program needs to format a string, you could add code to your program to do that, but writing code is hard. Alternatively, you could keep your program simple and instead connect to a service that provides string formatting for you. This has the added advantage that many programs can use this same service without reinventing the wheel.

Unfortunately, the microservice paradigm has largely avoided the C ecosystem; until now!

As a motivating example, we are going to build a TCP server that converts strings into uppercase. If a client connects and sends `Hello`, then our program will send `HELLO` back. This will serve as a very basic microservice. Of course, a real-world microservice might do something a bit more advanced (such as left-pad a string), but this to-uppercase service works well for our pedagogical purposes.

For our microservice to be useful, it does need to handle many simultaneous incoming connections. We again use `select()` to see which connections need to be serviced.

Our basic program flow looks like this:

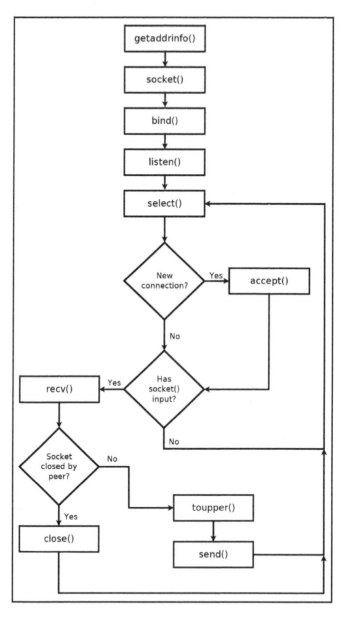

Like in `Chapter 2`, *Getting to Grips with Socket APIs*, our TCP server uses `getaddrinfo()` to obtain the local address to listen on. It creates a socket with `socket()`, uses `bind()` to associate the local address to the socket, and uses `listen()` to begin listening for new connections. Up until that point, it is essentially identical to our TCP server from `Chapter 2`, *Getting to Grips with Socket APIs*.

However, our next step is not to call `accept()` to wait for new connections. Instead, we call `select()`, which alerts us if a new connection is available or if any of our established connections have new data ready. Only when we know that a new connection is waiting do we call `accept()`. All established connections are put into an `fd_set`, which is passed to every subsequent `select()` call. In this same way, we know which connections would block on `recv()`, and we only service those connections that we know will not block.

When data is received by `recv()`, we run it through `toupper()` and return it to the client using `send()`.

This is a complicated program with several new concepts. Don't worry about understanding all the details right now. The flow is only intended to give you an overview of what to expect before we dive into the actual code.

TCP server code

Our TCP server code begins by including the needed headers, starting `main()`, and initializing Winsock. Refer to `Chapter 2`, *Getting to Grips with Socket APIs*, if this doesn't seem familiar:

```
/*tcp_serve_toupper.c*/

#include "chap03.h"
#include <ctype.h>

int main() {

#if defined(_WIN32)
    WSADATA d;
    if (WSAStartup(MAKEWORD(2, 2), &d)) {
        fprintf(stderr, "Failed to initialize.\n");
        return 1;
    }
#endif
```

We then get our local address, create our socket, and `bind()`. This is all done exactly as explained in Chapter 2, *Getting to Grips with Socket APIs*:

```
/*tcp_serve_toupper.c */

    printf("Configuring local address...\n");
    struct addrinfo hints;
    memset(&hints, 0, sizeof(hints));
    hints.ai_family = AF_INET;
    hints.ai_socktype = SOCK_STREAM;
    hints.ai_flags = AI_PASSIVE;

    struct addrinfo *bind_address;
    getaddrinfo(0, "8080", &hints, &bind_address);


    printf("Creating socket...\n");
    SOCKET socket_listen;
    socket_listen = socket(bind_address->ai_family,
            bind_address->ai_socktype, bind_address->ai_protocol);
    if (!ISVALIDSOCKET(socket_listen)) {
        fprintf(stderr, "socket() failed. (%d)\n", GETSOCKETERRNO());
        return 1;
    }
```

Note that we are going to listen on port 8080. You can, of course, change that. We're also doing an IPv4 server here. If you want to listen for connections on IPv6, then just change `AF_INET` to `AF_INET6`.

We then `bind()` our socket to the local address and have it enter a listening state. Again, this is done exactly as in Chapter 2, *Getting to Grips with Socket APIs*:

```
/*tcp_serve_toupper.c*/

    printf("Binding socket to local address...\n");
    if (bind(socket_listen,
                bind_address->ai_addr, bind_address->ai_addrlen)) {
        fprintf(stderr, "bind() failed. (%d)\n", GETSOCKETERRNO());
        return 1;
    }
    freeaddrinfo(bind_address);

    printf("Listening...\n");
    if (listen(socket_listen, 10) < 0) {
        fprintf(stderr, "listen() failed. (%d)\n", GETSOCKETERRNO());
        return 1;
    }
```

This is the point where we diverge from our earlier methods. We now define an `fd_set` structure that stores all of the active sockets. We also maintain a `max_socket` variable, which holds the largest socket descriptor. For now, we add only our listening socket to the set. Because it's the only socket, it must also be the largest, so we set `max_socket = socket_listen` too:

```
/*tcp_serve_toupper.c */

    fd_set master;
    FD_ZERO(&master);
    FD_SET(socket_listen, &master);
    SOCKET max_socket = socket_listen;
```

Later in the program, we will add new connections to `master` as they are established.

We then print a status message, enter the main loop, and set up our call to `select()`:

```
/*tcp_serve_toupper.c */

    printf("Waiting for connections...\n");

    while(1) {
        fd_set reads;
        reads = master;
        if (select(max_socket+1, &reads, 0, 0, 0) < 0) {
            fprintf(stderr, "select() failed. (%d)\n", GETSOCKETERRNO());
            return 1;
        }
```

This works by first copying our `fd_set master` into `reads`. Recall that `select()` modifies the set given to it. If we didn't copy `master`, we would lose its data.

We pass a timeout value of 0 (NULL) to `select()` so that it doesn't return until a socket in the `master` set is ready to be read from. At the beginning of our program, `master` only contains `socket_listen`, but as our program runs, we add each new connection to `master`.

We now loop through each possible socket and see whether it was flagged by `select()` as being ready. If a socket, X, was flagged by `select()`, then `FD_ISSET(X, &reads)` is true. Socket descriptors are positive integers, so we can try every possible socket descriptor up to `max_socket`. The basic structure of our loop is as follows:

```
/*tcp_serve_toupper.c */

    SOCKET i;
    for(i = 1; i <= max_socket; ++i) {
```

```
            if (FD_ISSET(i, &reads)) {
                //Handle socket
            }
        }
```

Remember, `FD_ISSET()` is only true for sockets that are ready to be read. In the case of `socket_listen`, this means that a new connection is ready to be established with `accept()`. For all other sockets, it means that data is ready to be read with `recv()`. We should first determine whether the current socket is the listening one or not. If it is, we call `accept()`. This code snippet and the one that follows replace the `//Handle socket` comment in the preceding code:

```
/*tcp_serve_toupper.c */

            if (i == socket_listen) {
                struct sockaddr_storage client_address;
                socklen_t client_len = sizeof(client_address);
                SOCKET socket_client = accept(socket_listen,
                        (struct sockaddr*) &client_address,
                        &client_len);
                if (!ISVALIDSOCKET(socket_client)) {
                    fprintf(stderr, "accept() failed. (%d)\n",
                            GETSOCKETERRNO());
                    return 1;
                }

                FD_SET(socket_client, &master);
                if (socket_client > max_socket)
                    max_socket = socket_client;

                char address_buffer[100];
                getnameinfo((struct sockaddr*)&client_address,
                        client_len,
                        address_buffer, sizeof(address_buffer), 0, 0,
                        NI_NUMERICHOST);
                printf("New connection from %s\n", address_buffer);
```

If the socket is `socket_listen`, then we `accept()` the connection much as we did in Chapter 2, *Getting to Grips with Socket APIs*. We use `FD_SET()` to add the new connection's socket to the `master` socket set. This allows us to monitor it with subsequent calls to `select()`. We also maintain `max_socket`. As a final step, this code prints out the client's address using `getnameinfo()`.

If the socket i is not `socket_listen`, then it is instead a request for an established connection. In this case, we need to read it with `recv()`, convert it into uppercase using the built-in `toupper()` function, and send the data back:

```
/*tcp_serve_toupper.c */

            } else {
                char read[1024];
                int bytes_received = recv(i, read, 1024, 0);
                if (bytes_received < 1) {
                    FD_CLR(i, &master);
                    CLOSESOCKET(i);
                    continue;
                }

                int j;
                for (j = 0; j < bytes_received; ++j)
                    read[j] = toupper(read[j]);
                send(i, read, bytes_received, 0);
            }
```

If the client has disconnected, then `recv()` returns a non-positive number. In this case, we remove that socket from the `master` socket set, and we also call `CLOSESOCKET()` on it to clean up.

Our program is now almost finished. We can end the if `FD_ISSET()` statement, end the for loop, end the `while` loop, close the listening socket, and clean up Winsock:

```
/*tcp_serve_toupper.c */

            } //if FD_ISSET
        } //for i to max_socket
    } //while(1)

    printf("Closing listening socket...\n");
    CLOSESOCKET(socket_listen);

#if defined(_WIN32)
    WSACleanup();
#endif

    printf("Finished.\n");
    return 0;
}
```

Our program is set up to continuously listen for connections, so the code after the end of the `while` loop will never run. Nevertheless, I believe it is still good practice to include it in case we program in functionality later to abort the `while` loop.

That's the complete to-uppercase microservice TCP server program. You can compile and run it on Linux and macOS like this:

```
gcc tcp_serve_toupper.c -o tcp_serve_toupper
./tcp_serve_toupper
```

Compiling and running on Windows with MinGW is done like this:

```
gcc tcp_serve_toupper.c -o tcp_serve_toupper.exe -lws2_32
tcp_serve_toupper.exe
```

You can abort the program's execution with *Ctrl + C*.

Once the program is running, I would suggest opening another terminal and running the `tcp_client` program from earlier to connect to it:

```
tcp_client 127.0.0.1 8080
```

Anything you type in `tcp_client` should be sent back as uppercase. Here's what this might look like:

As a test of the server program's functionality, try opening several additional terminals and connecting with `tcp_client`. Our server should be able to handle many simultaneous connections.

Also included with this chapter's code is `tcp_serve_toupper_fork.c`. This program only runs on Unix-based operating systems, but it performs the same functions as `tcp_serve_toupper.c` by using `fork()` instead of `select()`. The `fork()` function is commonly used by TCP servers, so I think it's helpful to be familiar with it.

Building a chat room

It is also possible, and common, to need to send data between connected clients. We can modify our `tcp_serve_toupper.c` program and make it a chat room pretty easily.

First, locate the following code in `tcp_serve_toupper.c`:

```
/*tcp_serve_toupper.c excerpt*/

                int j;
                for (j = 0; j < bytes_received; ++j)
                        read[j] = toupper(read[j]);
                send(i, read, bytes_received, 0);
```

Replace the preceding code with the following:

```
/*tcp_serve_chat.c excerpt*/

                SOCKET j;
                for (j = 1; j <= max_socket; ++j) {
                        if (FD_ISSET(j, &master)) {
                                if (j == socket_listen || j == i)
                                        continue;
                                else
                                        send(j, read, bytes_received, 0);
                        }
                }
```

This works by looping through all the sockets in the `master` set. For each socket, `j`, we check that it's not the listening socket and we check that it's not the same socket that sent the data in the first place. If it's not, we call `send()` to echo the received data to it.

You can compile and run this program in the same way as the previous one.

On Linux and macOS, this is done as follows:

```
gcc tcp_serve_chat.c -o tcp_serve_chat
./tcp_serve_chat
```

On Windows, this is done as follows:

```
gcc tcp_serve_chat.c -o tcp_serve_chat.exe -lws2_32
tcp_serve_chat.exe
```

You should open two or more additional windows and connect to it with the following code:

```
tcp_client 127.0.0.1 8080
```

Whatever you type in one of the tcp_client terminals get sent to all of the other connected terminals.

Here is an example of what this may look like:

In the preceding screenshot, I am running tcp_serve_chat in the upper-left terminal windows. The other three terminal windows are running tcp_client. As you can see, any text entered in one of the tcp_client windows is sent to the server, which relays it to the other two connected clients.

Blocking on send()

When we call send() with an amount of data, send() first copies this data into an outgoing buffer provided by the operating system. If we call send() when its outgoing buffer is already full, it blocks until its buffer has emptied enough to accept more of our data.

In some cases where send() would block, it instead returns without copying all of the data as requested. In this case, the return value of send() indicates how many bytes were actually copied. One example of this is if your program is blocking on send() and then receives a signal from the operating system. In these cases, it is up to the caller to try again with any remaining data.

In this chapter's *TCP server code* section, we ignored the possibility that send() could block or be interrupted. In a fully robust application, what we need to do is compare the return value from send() with the number of bytes that we tried to send. If the number of bytes actually sent is less than requested, we should use select() to determine when the socket is ready to accept new data, and then call send() with the remaining data. As you can imagine, this can become a bit complicated when keeping track of multiple sockets.

As the operating system usually provides a large enough outgoing buffer, we were able to avoid this possibility with our earlier server code. If we know that our server may try to send large amounts of data, we should certainly check for the return value from send().

The following code example assumes that buffer contains buffer_len bytes of data to send over a socket called peer_socket. This code blocks until we've sent all of buffer or an error (such as the peer disconnecting) occurs:

```
int begin = 0;
while (begin < buffer_len) {
    int sent = send(peer_socket, buffer + begin, buffer_len - begin, 0);
    if (sent == -1) {
        //Handle error
    }
    begin += sent;
}
```

If we are managing multiple sockets and don't want to block, then we should put all sockets with pending `send()` into an `fd_set` and pass it as the third parameter to `select()`. When `select()` signals on these sockets, then we know that they are ready to send more data.

Chapter 13, *Socket Programming Tips and Pitfalls*, addresses concerns regarding the `send()` function's blocking behavior in more detail.

TCP is a stream protocol

A common mistake beginners make is assuming that any data passed into `send()` can be read by `recv()` on the other end in the same amount. In reality, sending data is similar to writing and reading from a file. If we write 10 bytes to a file, followed by another 10 bytes, then the file has 20 bytes of data. If the file is to be read later, we could read 5 bytes and 15 bytes, or we could read all 20 bytes at once, and so on. In any case, we have no way of knowing that the file was written in two 10 byte chunks.

Using `send()` and `recv()` works the same way. If you `send()` 20 bytes, it's not possible to tell how many `recv()` calls these bytes are partitioned into. It is possible that one call to `recv()` could return all 20 bytes, but it is also possible that a first call to `recv()` returns 16 bytes and that a second call to `recv()` is needed to get the last 4 bytes.

This can make communication difficult. In many protocols, as we will see later in this book, it is important that received data be buffered up until enough of it has accumulated to warrant processing. We avoided this issue in this chapter with our to-uppercase sever by defining a protocol that operates just as well on 1 byte as it does on 100. This isn't true for most application protocols.

For a concrete example, imagine we wanted to make our `tcp_serve_toupper` server terminate if it received the `quit` command through a TCP socket. You could call `send(socket, "quit", 4, 0)` on the client and you may think that a call to `recv()` on the server would return `quit`. Indeed, in your testing, it is very likely to work that way. However, this behavior is not guaranteed. A call to `recv()` could just as likely return `qui`, and a second call to `recv()` may be required to receive the last `t`. If that is the case, consider how you would interpret whether a `quit` command has been received. The straightforward way to do it would be to buffer up data that's received from multiple `recv()` calls.

We will cover techniques for dealing with `recv()` buffering in `Section 2`, *An Overview of Application Layer Protocols*, of this book.

In contrast to TCP, UDP is not a stream protocol. With UDP, a packet is received with exactly the same contents as it was sent with. This can sometimes make handling UDP somewhat easier, as we will see in `Chapter 4`, *Establishing UDP Connections*.

Summary

TCP really serves as the backbone of the modern internet experience. TCP is used by HTTP, the protocol that powers websites, and by **Simple Mail Transfer Protocol** (**SMTP**), the protocol that powers email.

In this chapter, we saw that building a TCP client was fairly straightforward. The only really tricky part was having the client monitor for local terminal input while simultaneously monitoring for socket data. We were able to accomplish this with `select()` on Unix-based systems, but it was slightly trickier on Windows. Many real-world applications don't need to monitor terminal input, and so this step isn't always needed.

Building a TCP server that's suitable for many parallel connections wasn't much harder. Here, `select()` was extremely useful, as it allowed a straightforward way of monitoring the listening socket for new connections while also monitoring existing connections for new data.

We also touched briefly on some common pain points. TCP doesn't provide a native way to partition data. For more complicated protocols where this is needed, we have to buffer data from `recv()` until a suitable amount is available to interpret. For TCP peers that are handling large amounts of data, buffering to `send()` is also necessary.

The next chapter, `Chapter 4`, *Establishing UDP Connections*, is all about UDP, the counterpart to TCP. In some ways, UDP programming is simpler than TCP programming, but it is also very different.

Questions

Try answering these questions to test your knowledge on this chapter:

1. How can we tell whether the next call to `recv()` will block?
2. How can you ensure that `select()` doesn't block for longer than a specified time?
3. When we used our `tcp_client` program to connect to a web server, why did we need to send a blank line before the web server responded?
4. Does `send()` ever block?
5. How can we tell whether the socket has been disconnected by our peer?
6. Is data received by `recv()` always the same size as data sent with `send()`?
7. Consider the following code:

```
recv(socket_peer, buffer, 4096, 0);
printf(buffer);
```

   What is wrong with it?
   Also see what is wrong with this code:

```
recv(socket_peer, buffer, 4096, 0);
printf("%s", buffer);
```

The answers can be found in `Appendix A`, *Answers to Questions*.

Establishing UDP Connections

4

In this chapter, we will look at how to send and receive **User Datagram Protocol** (UDP) packets. UDP socket programming is very similar to **Transmission Control Protocol** (TCP) socket programming, so it is recommended that you read and understand Chapter 3, *An In-Depth Overview of TCP Connections*, before beginning this one.

The following topics are covered in this chapter:

- The differences between TCP socket programming and UDP socket programming
- The sendto() and recvfrom() functions
- How connect() works on a UDP socket
- Implementing a UDP server using only one socket
- Using select() to tell when a UDP socket has data ready

Technical requirements

The example programs in this chapter can be compiled with any modern C compiler. We recommend MinGW on Windows and GCC on Linux and macOS. See Appendix B, *Setting Up Your C Compiler On Windows*, Appendix C, *Setting Up Your C Compiler On Linux*, and Appendix D, *Setting Up Your C Compiler On macOS*, for compiler setup.

The code for this book can be found in this book's GitHub repository: `https://github.com/codeplea/Hands-On-Network-Programming-with-C`.

From the command line, you can download the code for this chapter with the following command:

```
git clone https://github.com/codeplea/Hands-On-Network-Programming-with-C
cd Hands-On-Network-Programming-with-C/chap04
```

Each example program in this chapter runs on Windows, Linux, and macOS. While compiling on Windows, each example program requires being linked with the Winsock library. This can be accomplished by passing the `-lws2_32` option to `gcc`.

We provide the exact commands that are needed to compile each example as they are introduced.

All of the example programs in this chapter require the same header files and C macros that we developed in Chapter 2, *Getting to Grips with Socket APIs*. For brevity, we put these statements in a separate header file, `chap04.h`, which we can include in each program. For an explanation of these statements, please refer to Chapter 2, *Getting to Grips with Socket APIs*.

The content of `chap04.h` is shown in the following code:

```c
#if defined(_WIN32)
#ifndef _WIN32_WINNT
#define _WIN32_WINNT 0x0600
#endif
#include <winsock2.h>
#include <ws2tcpip.h>
#pragma comment(lib, "ws2_32.lib")

#else
#include <sys/types.h>
#include <sys/socket.h>
#include <netinet/in.h>
#include <arpa/inet.h>
#include <netdb.h>
#include <unistd.h>
#include <errno.h>

#endif


#if defined(_WIN32)
#define ISVALIDSOCKET(s) ((s) != INVALID_SOCKET)
```

```
#define CLOSESOCKET(s) closesocket(s)
#define GETSOCKETERRNO() (WSAGetLastError())

#else
#define ISVALIDSOCKET(s) ((s) >= 0)
#define CLOSESOCKET(s) close(s)
#define SOCKET int
#define GETSOCKETERRNO() (errno)
#endif


#include <stdio.h>
#include <string.h>
```

How UDP sockets differ

The socket API for UDP sockets is only very slightly different than what we've already learned for TCP. In fact, they are similar enough that we can take the TCP client from the last chapter and turn it into a fully functional UDP client by changing only one line of code:

1. Take `tcp_client.c` from Chapter 3, *An In-Depth Overview of TCP Connections*, and find the following line of code:

    ```
    hints.ai_socktype = SOCK_STREAM;
    ```

2. Change the preceding code to the following:

    ```
    hints.ai_socktype = SOCK_DGRAM;
    ```

This modification is included in this chapter's code as `udp_client.c`.

You can recompile the program using the same commands as before, and you'll get a fully functional UDP client.

Unfortunately, changing the TCP servers of the previous chapters to UDP won't be as easy. TCP and UDP server code are different enough that a slightly different approach is needed.

Also, don't assume that because we had to change only one line of the code that the client behaves exactly the same way – this won't happen. The two programs are using a different protocol, after all.

Remember from Chapter 2, *Getting to Grips with Socket APIs* that UDP does not try to be a reliable protocol. Lost packets are not automatically re-transmitted, and packets may be received in a different order than they were sent. It is even possible for one packet to erroneously arrive twice! TCP attempts to solve all these problems, but UDP leaves you to your own devices.

Do you know what the best thing about a UDP joke is? I don't care if you get it or not.

Despite UDP's (lack of) reliability, it is still appropriate for many applications. Let's look at the methods that are used by UDP clients and servers.

UDP client methods

Sending data with TCP requires calling connect() to set the remote address and establish the TCP connection. Thus, we use send() with TCP sockets, as shown in the following code:

```
connect(tcp_socket, peer_address, peer_address_length);
send(tcp_socket, data, data_length, 0);
```

UDP is a connectionless protocol. Therefore, no connection is established before sending data. A UDP connection is never established. With UDP, data is simply sent and received. We can call connect() and then send(), as we mentioned previously, but the socket API provides an easier way for UDP sockets in the form of the sendto() function. It works like this:

```
sendto(udp_socket, data, data_length, 0,
        peer_address, peer_address_length);
```

connect() on a UDP socket works a bit differently. All connect() does with a UDP socket is associate a remote address. Thus, while connect() on a TCP socket involves a handshake for sending packets over the network, connect() on a UDP socket only stores an address locally.

So, a UDP client can be structured in two different ways, depending on whether you use connect(), send(), and recv(), or instead use sendto() and recvfrom().

The following diagram compares the program flow of a **TCP Client** to a **UDP Client** using either method:

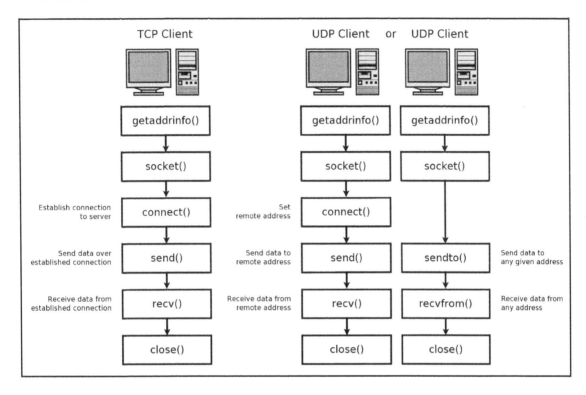

Note that, while using `connect()`, the **UDP Client** only receives data from the peer having the IP address and the port that is given to `connect()`. However, when not using `connect()`, the `recvfrom()` function returns data from any peer that addresses us! Of course, that peer would need to know our address and port. Unless we call `bind()`, our local address and port is assigned automatically by the operating system.

UDP server methods

Programming a UDP server is a bit different than TCP. TCP requires managing a socket for each peer connection. With UDP, our program only needs one socket. That one socket can communicate with any number of peers.

While the TCP program flow required us to use `listen()` and `accept()` to wait for and establish new connections, these functions are not used with UDP. Our UDP server simply binds to the local address, and then it can immediately start sending and receiving data.

The program flow of a **TCP Server** compared to a **UDP Server** is as follows:

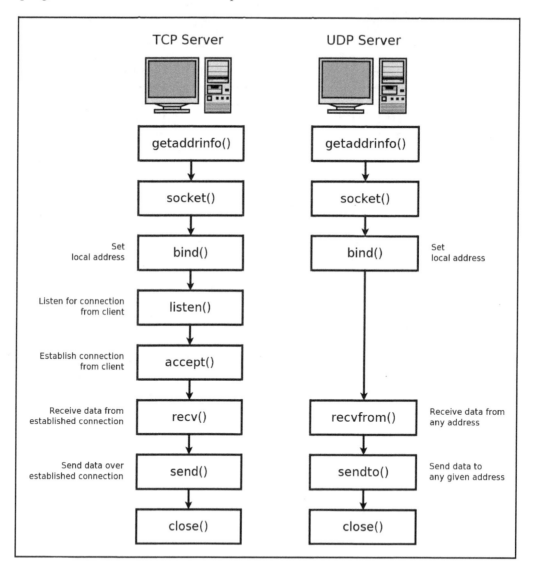

With either a TCP or UDP server, we use `select()` when we need to check/wait for incoming data. The difference is that a **TCP Server** using `select()` is likely monitoring many separate sockets, while a **UDP Server** often only needs to monitor one socket. If your program uses both TCP and UDP sockets, you can monitor them all with only one call to `select()`.

A first UDP client/server

To drive these points home, it will be useful to work through a full UDP client and UDP server program.

To keep things simple, we will create a UDP client program that simply sends the `Hello World` string to `127.0.0.1` on port `8080`. Our UDP server listens on `8080`. It prints any data it receives, along with the sender's address and port number.

We will begin by implementing the simple UDP server.

A simple UDP server

We will start with the server, since we already have a usable UDP client, that is, `udp_client.c`.

Like all of our networked programs, we will begin by including the necessary headers, starting with the `main()` function, and initializing Winsock as follows:

```
/*udp_recvfrom.c*/

#include "chap04.h"

int main() {

#if defined(_WIN32)
    WSADATA d;
    if (WSAStartup(MAKEWORD(2, 2), &d)) {
        fprintf(stderr, "Failed to initialize.\n");
        return 1;
    }
#endif
```

If you've been working through this book in order, this code should be very routine for you by now. If you haven't, then please refer to Chapter 2, *Getting to Grips with Socket APIs*.

Then, we must configure the local address that our server listens on. We use getaddrinfo() for this, as follows:

```
/*udp_recvfrom.c continued*/

    printf("Configuring local address...\n");
    struct addrinfo hints;
    memset(&hints, 0, sizeof(hints));
    hints.ai_family = AF_INET;
    hints.ai_socktype = SOCK_DGRAM;
    hints.ai_flags = AI_PASSIVE;

    struct addrinfo *bind_address;
    getaddrinfo(0, "8080", &hints, &bind_address);
```

This differs only slightly from how we've done it before. Notably, we set hints.ai_socktype = SOCK_DGRAM. Recall that SOCK_STREAM was used there for TCP connections. We are still setting hints.ai_family = AF_INET here. This makes our server listen for IPv4 connections. We could change that to AF_INET6 to make our server listen for IPv6 connections instead.

After we have our local address information, we can create the socket, as follows:

```
/*udp_recvfrom.c continued*/

    printf("Creating socket...\n");
    SOCKET socket_listen;
    socket_listen = socket(bind_address->ai_family,
            bind_address->ai_socktype, bind_address->ai_protocol);
    if (!ISVALIDSOCKET(socket_listen)) {
        fprintf(stderr, "socket() failed. (%d)\n", GETSOCKETERRNO());
        return 1;
    }
```

This code is exactly the same as in the TCP case. The call to socket() uses our address information from getaddrinfo() to create the proper type of socket.

We must then bind the new socket to the local address that we got from getaddrinfo(). This is as follows:

```
/*udp_recvfrom.c continued*/

    printf("Binding socket to local address...\n");
    if (bind(socket_listen,
```

```
                    bind_address->ai_addr, bind_address->ai_addrlen)) {
        fprintf(stderr, "bind() failed. (%d)\n", GETSOCKETERRNO());
        return 1;
    }
    freeaddrinfo(bind_address);
```

Again, that code is exactly the same as in the TCP case.

Here is where the UDP server diverges from the TCP server. Once the local address is bound, we can simply start to receive data. There is no need to call `listen()` or `accept()`. We listen for incoming data using `recvfrom()`, as shown here:

```
/*udp_recvfrom.c continued*/

    struct sockaddr_storage client_address;
    socklen_t client_len = sizeof(client_address);
    char read[1024];
    int bytes_received = recvfrom(socket_listen,
            read, 1024,
            0,
            (struct sockaddr*) &client_address, &client_len);
```

In the previous code, we created a `struct sockaddr_storage` to store the client's address. We also defined `socklen_t client_len` to hold the address size. This keeps our code robust in the case that we change it from IPv4 to IPv6. Finally, we created a buffer, `char read[1024]`, to store incoming data.

`recvfrom()` is used in a similar manner to `recv()`, except that it returns the sender's address, as well as the received data. You can think of `recvfrom()` as a combination of the TCP server `accept()` and `recv()`.

Once we've received data, we can print it out. Keep in mind that the data may not be null terminated. It can be safely printed with the `%.*s printf()` format specifier, as shown in the following code:

```
/*udp_recvfrom.c continued*/

    printf("Received (%d bytes): %.*s\n",
            bytes_received, bytes_received, read);
```

It may also be useful to print the sender's address and port number. We can use the `getnameinfo()` function to convert this data into a printable string, as shown in the following code:

```
/*udp_recvfrom.c continued*/

    printf("Remote address is: ");
    char address_buffer[100];
    char service_buffer[100];
    getnameinfo(((struct sockaddr*)&client_address),
            client_len,
            address_buffer, sizeof(address_buffer),
            service_buffer, sizeof(service_buffer),
            NI_NUMERICHOST | NI_NUMERICSERV);
    printf("%s %s\n", address_buffer, service_buffer);
```

The last argument to `getnameinfo()` (NI_NUMERICHOST | NI_NUMERICSERV) tells `getnameinfo()` that we want both the client address and port number to be in numeric form. Without this, it would attempt to return a hostname or protocol name if the port number matches a known protocol. If you do want a protocol name, pass in the NI_DGRAM flag to tell `getnameinfo()` that you're working on a UDP port. This is important for the few protocols that have different ports for TCP and UDP.

It's also worth noting that the client will rarely set its local port number explicitly. So, the port number returned here by `getnameinfo()` is likely to be a high number that's chosen randomly by the client's operating system. Even if the client did set its local port number, the port number we can see here might have been changed by **network address translation** (**NAT**).

In any case, if our server were to send data back, it would need to send it to the address and port stored in `client_address`. This would be done by passing `client_address` to `sendto()`.

Once the data has been received, we'll end our simple UDP server by closing the connection, cleaning up Winsock, and ending the program:

```
/*udp_recvfrom.c continued*/

    CLOSESOCKET(socket_listen);

#if defined(_WIN32)
    WSACleanup();
#endif

    printf("Finished.\n");
```

```
    return 0;
}
```

You can compile and run `udp_recvfrom.c` on Linux and macOS by using the following commands:

```
gcc udp_recvfrom.c -o udp_recvfrom
./udp_recvfrom
```

Compiling and running on Windows with MinGW is done as follows:

```
gcc udp_recvfrom.c -o udp_recvfrom.exe -lws2_32
udp_recvfrom.exe
```

While running, it simply waits for an incoming connection:

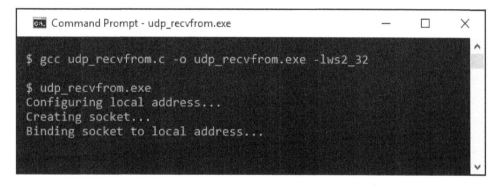

You could use `udp_client` to connect to `udp_recvfrom` for testing, or you can implement `udp_sendto`, which we will do next.

A simple UDP client

Although we've already shown a fairly full-featured UDP client, `udp_client.c`, it is worthwhile building a very simple UDP client. This client shows only the minimal required steps to get a working UDP client, and it uses `sendto()` instead of `send()`.

Let's begin the same way we begin each program, by including the necessary headers, starting `main()`, and initializing Winsock, as follows:

```
/*udp_sendto.c*/

#include "chap04.h"

int main() {

#if defined(_WIN32)
    WSADATA d;
    if (WSAStartup(MAKEWORD(2, 2), &d)) {
        fprintf(stderr, "Failed to initialize.\n");
        return 1;
    }
#endif
```

We then configure the remote address using `getaddrinfo()`. In this minimal example, we use `127.0.0.1` as the remote address and `8080` as the remote port. This means that it connects to the UDP server only if it's running on the same computer.

Here is how the remote address is configured:

```
/*udp_sendto.c continued*/

    printf("Configuring remote address...\n");
    struct addrinfo hints;
    memset(&hints, 0, sizeof(hints));
    hints.ai_socktype = SOCK_DGRAM;
    struct addrinfo *peer_address;
    if (getaddrinfo("127.0.0.1", "8080", &hints, &peer_address)) {
        fprintf(stderr, "getaddrinfo() failed. (%d)\n", GETSOCKETERRNO());
        return 1;
    }
```

Notice that we hardcoded `127.0.0.1` and `8080` into the call to `getaddrinfo()`. Also, notice that we've set `hints.ai_socktype = SOCK_DGRAM`. This tells `getaddrinfo()` that we are connecting over UDP. Notice that we did not set `AF_INET` or `AF_INET6`. This allows `getaddrinfo()` to return the appropriate address for IPv4 or IPv6. In this case, it is IPv4 because the address, `127.0.0.1`, is an IPv4 address. We will cover `getaddrinfo()` in more detail in `Chapter 5`, *Hostname Resolution and DNS*.

We can print the configured address using `getnameinfo()`. The call to `getnameinfo()` is the same as in the previous UDP server, `udp_recvfrom.c`. It works as follows:

```
/*udp_sendto.c continued*/

    printf("Remote address is: ");
    char address_buffer[100];
    char service_buffer[100];
    getnameinfo(peer_address->ai_addr, peer_address->ai_addrlen,
            address_buffer, sizeof(address_buffer),
            service_buffer, sizeof(service_buffer),
            NI_NUMERICHOST  | NI_NUMERICSERV);
    printf("%s %s\n", address_buffer, service_buffer);
```

Now that we've stored the remote address, we are ready to create our socket with a call to `socket()`. We pass in fields from `peer_address` to create the appropriate socket type. The code for this is as follows:

```
/*udp_sendto.c continued*/

    printf("Creating socket...\n");
    SOCKET socket_peer;
    socket_peer = socket(peer_address->ai_family,
            peer_address->ai_socktype, peer_address->ai_protocol);
    if (!ISVALIDSOCKET(socket_peer)) {
        fprintf(stderr, "socket() failed. (%d)\n", GETSOCKETERRNO());
        return 1;
    }
```

Once the socket has been created, we can go straight to sending data with `sendto()`. There is no need to call `connect()`. Here is the code to send `Hello World` to our UDP server:

```
/*udp_sendto.c continued*/

    const char *message = "Hello World";
    printf("Sending: %s\n", message);
    int bytes_sent = sendto(socket_peer,
            message, strlen(message),
            0,
            peer_address->ai_addr, peer_address->ai_addrlen);
    printf("Sent %d bytes.\n", bytes_sent);
```

Notice that `sendto()` is much like `send()`, except that we need to pass in an address as the last parameter.

It is also worth noting that we do not get an error back if sending fails. send() simply tries to send a message, but if it gets lost or is misdelivered along the way, there is nothing we can do about it. If the message is important, it is up to the application protocol to implement the corrective action.

After we've sent our data, we could reuse the same socket to send data to another address (as long as it's the same type of address, which is IPv4 in this case). We could also try to receive a reply from the UDP server by calling recvfrom(). Note that if we did call recvfrom() here, we could get data from anybody that sends to us – not necessarily the server we just transmitted to.

When we sent our data, our socket was assigned with a temporary local port number by the operating system. This local port number is called the **ephemeral port number**. From then on, our socket is essentially listening for a reply on this local port. If the local port is important, you can use bind() to associate a specific port before calling send().

If multiple applications on the same system are connecting to a remote server at the same port, the operating system uses the local ephemeral port number to keep replies separate. Without this, it wouldn't be possible to know which application should receive which reply.

We'll end our example program by freeing the memory for peer_address, closing the socket, cleaning up Winsock, and finishing main(), as follows:

```
/*udp_sendto.c continued*/

    freeaddrinfo(peer_address);
    CLOSESOCKET(socket_peer);

#if defined(_WIN32)
    WSACleanup();
#endif

    printf("Finished.\n");
    return 0;
}
```

You can compile udp_sendto.c on Linux and macOS using the following command:

```
gcc udp_sendto.c -o udp_sendto
```

Compiling on Windows with MinGW is done in the following way:

```
gcc udp_sendto.c -o udp_sendto.exe -lws2_32
```

To test it out, first, start udp_recvfrom in a separate terminal. With udp_recvfrom already running, you can start udp_sendto. It should look as follows:

If no server is running on port 8080, udp_sendto still produces the same output. udp_sendto doesn't know that the packet was not delivered.

A UDP server

It will be useful to look at a UDP server that's been designed to service many connections. Fortunately for us, the UDP socket API makes this very easy.

We will take the motivating example from our last chapter, which was to provide a service that converts all text into uppercase. This is useful because you can directly compare the UDP code from here to the TCP server code from Chapter 3, *An In-Depth Overview of TCP Connections*.

Our server begins by setting up the socket and binding to our local address. It then waits to receive data. Once it has received a data string, it converts the string into all uppercase and sends it back.

The program flow looks as follows:

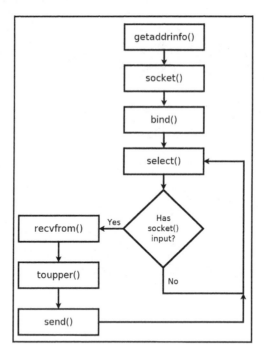

If you compare the flow of this program to the TCP server from the last chapter (Chapter 3, *An In-Depth Overview of TCP Connections*), you will find that it's much simpler. With TCP, we had to use `listen()` and `accept()`. With UDP, we skip those calls and go straight into receiving data with `recvfrom()`. With our TCP server, we had to monitor a listening socket for new connections while simultaneously monitoring an additional socket for each connected client. Our UDP server only uses one socket, so there is much less to keep track of.

Our UDP server program begins by including the necessary headers, starting the `main()` function, and initializing Winsock, as follows:

```
/*udp_serve_toupper.c*/

#include "chap04.h"
#include <ctype.h>

int main() {

#if defined(_WIN32)
    WSADATA d;
    if (WSAStartup(MAKEWORD(2, 2), &d)) {
        fprintf(stderr, "Failed to initialize.\n");
        return 1;
    }
#endif
```

We then find our local address that we should listen on, create the socket, and bind to it. This is all exactly the same as in our earlier server, `udp_recvfrom.c`. The only difference between this code and the TCP servers in Chapter 3, *An In-Depth Overview of TCP Connections*, is that we use `SOCK_DGRAM` instead of `SOCK_STREAM`. Recall that `SOCK_DGRAM` specifies that we want a UDP socket.

Here is the code for setting the address and creating a new socket:

```
/*udp_serve_toupper.c continued*/

    printf("Configuring local address...\n");
    struct addrinfo hints;
    memset(&hints, 0, sizeof(hints));
    hints.ai_family = AF_INET;
    hints.ai_socktype = SOCK_DGRAM;
    hints.ai_flags = AI_PASSIVE;

    struct addrinfo *bind_address;
    getaddrinfo(0, "8080", &hints, &bind_address);
```

```
    printf("Creating socket...\n");
    SOCKET socket_listen;
    socket_listen = socket(bind_address->ai_family,
            bind_address->ai_socktype, bind_address->ai_protocol);
    if (!ISVALIDSOCKET(socket_listen)) {
        fprintf(stderr, "socket() failed. (%d)\n", GETSOCKETERRNO());
        return 1;
    }
```

Binding the new socket to the local address is done as follows:

```
/*udp_serve_toupper.c continued*/

    printf("Binding socket to local address...\n");
    if (bind(socket_listen,
                bind_address->ai_addr, bind_address->ai_addrlen)) {
        fprintf(stderr, "bind() failed. (%d)\n", GETSOCKETERRNO());
        return 1;
    }
    freeaddrinfo(bind_address);
```

Because our server uses select(), we need to create a new fd_set to store our listening socket. We zero the set using FD_ZERO(), and then add our socket to this set using FD_SET(). We also maintain the maximum socket in the set using max_socket:

```
/*udp_serve_toupper.c continued*/

    fd_set master;
    FD_ZERO(&master);
    FD_SET(socket_listen, &master);
    SOCKET max_socket = socket_listen;

    printf("Waiting for connections...\n");
```

Note that we don't really have to use select() for this program, and omitting it would make the program simpler (see udp_server_toupper_simple.c). However, we are going to use select() because it makes our code more flexible. We could easily add in an additional socket (if we needed to listen on multiple ports, for example), and we could add in a select() timeout if our program needs to perform other functions. Of course, our program doesn't do those things, so we don't need select(), but I think that most programs do, so we will show it that way.

Now, we are ready for the main loop. It copies the socket set into a new variable, `reads`, and then uses `select()` to wait until our socket is ready to read from. Recall that we could pass in a timeout value as the last parameter to `select()` if we want to set a maximum waiting time for the next read. Refer to Chapter 3, *An In-Depth Overview of TCP Connections*, the *Synchronous multiplexing with select()* section, for more information on `select()`.

Once `select()` returns, we use `FD_ISSET()` to tell if our particular socket, `socket_listen`, is ready to be read from. If we had additional sockets, we would need to use `FD_ISSET()` for each socket.

If `FD_ISSET()` returns true, we read from the socket using `recvfrom()`. `recvfrom()` gives us the sender's address, so we must first allocate a variable to hold the address, that is, `client_address`. Once we've read a string from the socket using `recvfrom()`, we convert the string into uppercase using the C `toupper()` function. We then send the modified text back to the sender using `sendto()`. Note that the last two parameters to `sendto()` are the client's addresses that we got from `recvfrom()`.

The main program loop can be seen in the following code:

```
/*udp_serve_toupper.c continued*/

    while(1) {
        fd_set reads;
        reads = master;
        if (select(max_socket+1, &reads, 0, 0, 0) < 0) {
            fprintf(stderr, "select() failed. (%d)\n", GETSOCKETERRNO());
            return 1;
        }

        if (FD_ISSET(socket_listen, &reads)) {
            struct sockaddr_storage client_address;
            socklen_t client_len = sizeof(client_address);

            char read[1024];
            int bytes_received = recvfrom(socket_listen, read, 1024, 0,
                    (struct sockaddr *)&client_address, &client_len);
            if (bytes_received < 1) {
                fprintf(stderr, "connection closed. (%d)\n",
                        GETSOCKETERRNO());
                return 1;
            }

            int j;
            for (j = 0; j < bytes_received; ++j)
                read[j] = toupper(read[j]);
```

```
        sendto(socket_listen, read, bytes_received, 0,
            (struct sockaddr*)&client_address, client_len);

    } //if FD_ISSET
} //while(1)
```

We can then close the socket, clean up Winsock, and terminate the program. Note that this code never runs, because the main loop never terminates. We include this code anyway as good practice; in case the program is adapted in the future to have an exit function.

The cleanup code is as follows:

```
/*udp_serve_toupper.c continued*/

    printf("Closing listening socket...\n");
    CLOSESOCKET(socket_listen);

#if defined(_WIN32)
    WSACleanup();
#endif

    printf("Finished.\n");

    return 0;
}
```

That's our complete UDP server program. You can compile and run it on Linux and macOS as follows:

```
gcc udp_serve_toupper.c -o udp_serve_toupper
./udp_serve_toupper
```

Compiling and running on Windows with MinGW is done in the following manner:

```
gcc udp_serve_toupper.c -o udp_serve_toupper.exe -lws2_32
udp_serve_toupper.exe
```

You can abort the program's execution with *Ctrl + C*.

Once the program is running, you should open another terminal window and run the udp_client program from earlier to connect to it, as follows:

```
udp_client 127.0.0.1 8080
```

Anything you type in `udp_client` should be sent back to it in uppercase. Here's what that might look like:

You may also want to try opening additional terminal windows and connecting with `udp_client`.

See `udp_serve_toupper_simple.c` for an implementation that doesn't use `select()`, but manages to work just as well anyway.

Summary

In this chapter, we saw that programming with UDP sockets is somewhat easier than with TCP sockets. We learned that UDP sockets don't need the `listen()`, `accept()`, or `connect()` function calls. This is mostly because `sendto()` and `recvfrom()` deal with the addresses directly. For more complicated programs, we can still use the `select()` function to see which sockets are ready for I/O.

We also saw that UDP sockets are connectionless. This is in contrast to connection-oriented TCP sockets. With TCP, we had to establish a connection before sending data, but with UDP, we simply send individual packets directly to a destination address. This keeps UDP socket programming simple, but it can complicate application protocol design, and UDP does not automatically retry communication failures or ensure that packets arrive in order.

The next chapter, `Chapter 5`, *Hostname Resolution and DNS*, is all about hostnames. Hostnames are resolved using the DNS protocol, which works over UDP. Move on to `Chapter 5`, *Hostname Resolution and DNS*, to learn about implementing a real-world UDP protocol.

Questions

Try answering these questions to test your knowledge of this chapter:

1. How do `sendto()` and `recvfrom()` differ from `send()` and `recv()`?
2. Can `send()` and `recv()` be used on UDP sockets?
3. What does `connect()` do in the case of a UDP socket?
4. What makes multiplexing with UDP easier than with TCP?
5. What are the downsides to UDP when compared to TCP?
6. Can the same program use UDP and TCP?

The answers can be found in `Appendix A`, *Answers to Questions*.

Hostname Resolution and DNS

5

Hostname resolution is a vital part of network programming. It allows us to use simple names, such as `www.example.com`, instead of tedious addresses such as `::ffff:192.168.212.115`. The mechanism that allows us to resolve hostnames into IP addresses and IP addresses into hostnames is the **Domain Name System** (**DNS**).

In this chapter, we begin by covering the built-in `getaddrinfo()` and `getnameinfo()` socket functions in more depth. Later, we will build a program that does DNS queries using **User Datagram Protocol** (**UDP**) from scratch.

We will cover the following topics in this chapter:

- How DNS works
- Common DNS record types
- The `getaddrinfo()` and `getnameinfo()` functions
- DNS query data structures
- DNS UDP protocol
- DNS TCP fallback
- Implementing a DNS query program

Technical requirements

The example programs from this chapter can be compiled with any modern C compiler. We recommend MinGW on Windows and GCC on Linux and macOS. See Appendices B, C, and D for compiler setup.

The code for this book can be found at `https://github.com/codeplea/Hands-On-Network-Programming-with-C`.

From the command line, you can download the code for this chapter with the following command:

```
git clone https://github.com/codeplea/Hands-On-Network-Programming-with-C
cd Hands-On-Network-Programming-with-C/chap05
```

Each example program in this chapter runs on Windows, Linux, and macOS. When compiling on Windows, each example program should be linked with the Winsock library. This can be accomplished by passing the -lws2_32 option to gcc.

We'll provide the exact commands needed to compile each example as they are introduced.

All of the example programs in this chapter require the same header files and C macros which we developed in Chapter 2, *Getting to Grips with Socket APIs*. For brevity, we put these statements in a separate header file, chap05.h, which we can include in each program. For an explanation of these statements, please refer to Chapter 2, *Getting to Grips with Socket APIs*.

The contents of chap05.h are as follows:

```
/*chap05.h*/

#if defined(_WIN32)
#ifndef _WIN32_WINNT
#define _WIN32_WINNT 0x0600
#endif
#include <winsock2.h>
#include <ws2tcpip.h>
#pragma comment(lib, "ws2_32.lib")

#else
#include <sys/types.h>
#include <sys/socket.h>
#include <netinet/in.h>
#include <arpa/inet.h>
#include <netdb.h>
#include <unistd.h>
#include <errno.h>

#endif


#if defined(_WIN32)
#define ISVALIDSOCKET(s) ((s) != INVALID_SOCKET)
#define CLOSESOCKET(s) closesocket(s)
#define GETSOCKETERRNO() (WSAGetLastError())
```

```
#else
#define ISVALIDSOCKET(s) ((s) >= 0)
#define CLOSESOCKET(s) close(s)
#define SOCKET int
#define GETSOCKETERRNO() (errno)
#endif


#include <stdio.h>
#include <stdlib.h>
#include <string.h>
```

With the `chap05.h` header in place, writing portable network programs is much easier. Let's continue now with an explanation of how DNS works, and then we will move on to the actual example programs.

How hostname resolution works

The DNS is used to assign names to computers and systems connected to the internet. Similar to how a phone book can be used to link a phone number to a name, the DNS allows us to link a hostname to an IP address.

When your program needs to connect to a remote computer, such as `www.example.com`, it first needs to find the IP address for `www.example.com`. In this book so far, we have been using the built-in `getaddrinfo()` function for this purpose. When you call `getaddrinfo()`, your operating system goes through a number of steps to resolve the domain name.

First, your operating system checks whether it already knows the IP address for `www.example.com`. If you have used that hostname recently, the OS is allowed to remember it in a local cache for a time. This time is referred to as **time-to-live** (**TTL**) and is set by the DNS server responsible for the hostname.

If the hostname is not found in the local cache, then your operating system will need to query a DNS server. This DNS server is usually provided by your **Internet Service Provider** (**ISP**), but there are also many publicly-available DNS servers. When the DNS server receives a query, it also checks its local cache. This is useful because numerous systems may be relying on one DNS server. If a DNS server receives 1,000 requests for `gmail.com` in one minute, it only needs to resolve the hostname the first time. For the other 999 requests, it can simply return the memorized answer from its local cache.

If the DNS server doesn't have the requested DNS record in its cache, then it needs to query other DNS servers until it connects directly to the DNS server responsible for the target system. Here's an example query resolution broken down step-wise:

Client A's DNS server is trying to resolve `www.example.com` as follows:

1. It first connects to a root DNS server and asks for `www.example.com`.
2. The root DNS server directs it to ask the `.com` server.
3. Our client's DNS server then connects to the server responsible for `.com` and asks for `www.example.com`.
4. The `.com` DNS server gives our server the address of another server – the `example.com` DNS server.
5. Our DNS server finally connects to that server and asks about the record for `www.example.com`.
6. The `example.com` server then shares the address of `www.example.com`.
7. Our client's DNS server relays it back to our client.

The following diagram illustrates this visually:

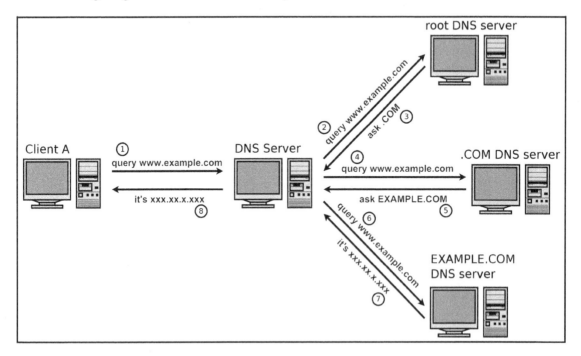

In this example, you can see that resolving `www.example.com` involved sending eight messages. It's possible that the lookup could have taken even longer. This is why it's imperative for DNS servers to implement caching. Let's assume that **Client B** tries the same query shortly after **Client A**; it's likely that the DNS server would have that value cached:

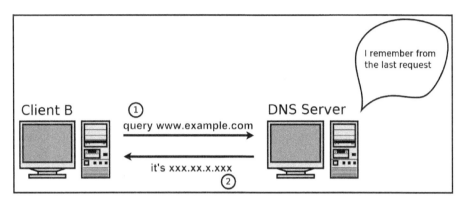

Of course, if a program on **Client A** wants to resolve `www.example.com` again, it's likely that it won't have to contact the DNS server at all – the operating system running on **Client A** should have cached the result.

On Windows, you can show your local DNS cache with the following command:

```
ipconfig /displaydns
```

On Linux or macOS, the command to show your local DNS varies depending on your exact system setup.

One downside of setting a large TTL value for a domain record is that you have to wait at least that long to be sure that all clients are using the new record and not just retrieving the old cached record.

In addition to DNS records that link a hostname to an IP address, there are other DNS record types for various purposes. We'll review some of these in the next section.

DNS record types

The DNS has five main types of records—A, AAAA, MX, TXT, CNAME, and * (ALL/ANY).

As we have learned, the DNS's primary purpose is to translate hostnames into IP addresses. This is done with two record types – type A and type AAAA. These records work in the same way, but A records return an IPv4 address, while AAAA records return an IPv6 address.

The MX record type is used to return mail server information. For example, if you wanted to send an email to larry@example.com, then the MX record(s) for example.com would indicate which mail server(s) receives emails for that domain.

TXT records can be used to store arbitrary information for a hostname. In practice, these are sometimes set to prove ownership of a domain name or to publish email sending guidelines. The **Sender Policy Framework (SPF)** standard uses TXT records to declare which systems are allowed to send mail for a given domain name. You can read more about SPF at http://www.openspf.org/.

CNAME records can be used to provide an alias for a given name. For example, many websites are accessible at both their root domain name, for example, example.com, and at the www subdomain. If example.com and www.example.com should point to the same address, then an A and an AAAA record can be added for example.com, while a CNAME record can be added for www.example.com pointing to example.com. Note that DNS clients don't query for CNAME records directly; instead, a client would ask for the A or AAAA record for www.example.com and the DNS server would reply with the CNAME record pointing to example.com. The DNS client would then continue the query using example.com.

When doing a DNS query, there is also a pseudo-record type called * or ALL or ANY. If this record is requested from a DNS server, then the DNS server returns all known record types for the current query. Note that a DNS server is allowed to respond with only the records in its cache, and this query is not guaranteed (or even likely) to actually get all of the records for the requested domain.

When sending a DNS query, each record type has an associated type ID. The IDs for the records discussed so far are as follows:

Record Type	Type ID (decimal)	Description
A	1	IPv4 address record
AAAA	28	IPv6 address record
MX	15	Mail exchange record
TXT	16	Text record
CNAME	5	Canonical name
*	255	All cached records

There are many other record types in use. Please see the *Further reading* section at the end of this chapter for more information.

It should be noted that one hostname may be associated with multiple records of the same type. For example, `example.com` could have several A records, each with a different IPv4 address. This is useful if multiple servers can provide the same service.

One other aspect of the DNS protocol worth mentioning is security. Let's look at that now.

DNS security

While most web traffic and email are encrypted today, DNS is still widely used in an unsecured manner. Protocols do exist to provide security for DNS, but they are not widely adopted yet. This will hopefully change in the near future.

Domain Name System Security Extensions (**DNSSEC**) are DNS extensions that provide data authentication. This authentication allows a DNS client to know that a given DNS reply is authentic, but it does not protect against eavesdropping.

DNS over HTTPS (**DoH**) is a protocol that provides name resolution over HTTPS. HTTPS provides strong security guarantees, including resistance to interception. We cover HTTPS in `Chapter 9`, *Loading Secure Web Pages with HTTPS and OpenSSL*, and `Chapter 10`, *Implementing a Secure Web Server*.

What are the implications of using insecure DNS? First, if DNS is not authenticated, then it could allow an attacker to lie about a domain name's IP address. This could trick a victim into connecting to a server that they think is `example.com`, but, in reality, is a malicious server controlled by the attacker at a different IP address. If the user is connecting via a secure protocol, such as HTTPS, then this attack will fail. HTTPS provides additional authentication to prove server identity. However, if the user connects with an insecure protocol, such as HTTP, then the DNS attack could successfully trick the victim into connecting to the wrong server.

If DNS is authenticated, then these hijacking attacks are prevented. However, without encryption, DNS queries are still susceptible to eavesdropping. This could potentially give an eavesdropper insight into which websites you're visiting and other servers that you are connecting to (for example, which email server you use). This doesn't let an attacker know what you are doing on each website. For example, if you do a DNS query for `example.com`, an attacker would know that you planned to visit `example.com`, but the attacker would not be able to determine which resources you requested from `example.com` – assuming that you use a secure protocol (for example, HTTPS) to retrieve those resources. An attacker with the ability to eavesdrop would be able to see that you established a connection to the IP address of `example.com` in any case, so them knowing that you performed a DNS lookup beforehand is not much extra information.

With some security discussion out of the way, let's look at how to do actual DNS lookups. Winsock and Berkeley sockets provide a simple function to do address lookup, called getaddrinfo(), which we've used in the previous chapters of this book. We will start with this in the next section.

Name/address translation functions

It is common for networked programs to need to translate text-based representatives of an address or hostname into an address structure required by the socket programming API. The common function we've been using is getaddrinfo(). It is a useful function because it is highly portable (available on Windows, Linux, and macOS), and it works for both IPv4 and IPv6 addresses.

It is also common to need to convert a binary address back into a text format. We use getnameinfo() for this.

Using getaddrinfo()

Although we've been using getaddrinfo() in previous chapters, we'll discuss it in more detail here.

The declaration for getaddrinfo() is shown in the following code:

```
int getaddrinfo(const char *node,
                const char *service,
                const struct addrinfo *hints,
                struct addrinfo **res);
```

The preceding code snippet is explained as follows:

- node specifies a hostname or address as a string. Valid examples could be example.com, 192.168.1.1, or ::1.
- service specifies a service or port number as a string. Valid examples could be http or 80. Alternately, the null pointer can be passed in for service, in which case, the resulting address is set to port 0.

- `hints` is a pointer to a `struct addrinfo`, which specifies options for selecting the address. The `addrinfo` structure has the following fields:

```
struct addrinfo {
    int               ai_flags;
    int               ai_family;
    int               ai_socktype;
    int               ai_protocol;
    socklen_t         ai_addrlen;
    struct sockaddr  *ai_addr;
    char             *ai_canonname;
    struct addrinfo  *ai_next;
};
```

You should not assume that the fields are stored in the order listed in the previous code, or that additional fields aren't present. There is some variation between operating systems.

The call to `getaddrinfo()` looks at only four fields in `*hints`. The rest of the structure should be zeroed-out. The relevant fields are `ai_family`, `ai_socktype`, `ai_protocol`, and `ai_flags`:

- `ai_family` specifies the desired address family. It can be `AF_INET` for IPv4, `AF_INET6` for IPv6, or `AF_UNSPEC` for any address family. `AF_UNSPEC` is defined as 0.

- `ai_socktype` could be `SOCK_STREAM` for TCP (see Chapter 3, *An In-Depth Overview of TCP Connections*), or `SOCK_DGRAM` for UDP (see Chapter 4, *Establishing UDP Connections*). Setting `ai_socktype` to 0 indicates that the address could be used for either.

- `ai_protocol` should be left to 0. Strictly speaking, TCP isn't the only streaming protocol supported by the socket interface, and UDP isn't the only datagram protocol supported. `ai_protocol` is used to disambiguate, but it's not needed for our purposes.

- `ai_flags` specifies additional options about how `getaddrinfo()` should work. Multiple flags can be used by bitwise OR-ing them together. In C, the bitwise OR operator uses the pipe symbol, `|`. So, bitwise OR-ing two flags together would use the `(flag_one | flag_two)` code.

Common flags you may use for the `ai_flags` field are:

- `AI_NUMERICHOST` can be used to prevent name lookups. In this case, `getaddrinfo()` would expect `node` to be an address such as `127.0.0.1` and not a hostname such as `example.com`. `AI_NUMERICHOST` can be useful because it prevents `getaddrinfo()` from doing a DNS record lookup, which can be slow.
- `AI_NUMERICSERV` can be used to only accept port numbers for the `service` argument. If used, this flag causes `getaddrinof()` to reject service names.
- `AI_ALL` can be used to request both an IPv4 and IPv6 address. The declaration for `AI_ALL` seems to be missing on some Windows setups. It can be defined as `0x0100` on those platforms.
- `AI_ADDRCONFIG` forces `getaddrinfo()` to only return addresses that match the family type of a configured interface on the local machine. For example, if your machine is IPv4 only, then using `AI_ADDRCONFIG | AI_ALL` prevents IPv6 addresses from being returned. It is usually a good idea to use this flag if you plan to connect a socket to the address returned by `getaddrinfo()`.
- `AI_PASSIVE` can be used with `node = 0` to request the *wildcard address*. This is the local address that accepts connections on any of the host's network addresses. It is used on servers with `bind()`. If `node` is not 0, then `AI_PASSIVE` has no effect. See `Chapter 3`, *An In-Depth Overview of TCP Connections* for example usage.

All other fields in `hints` should be set to 0. You can also pass in 0 for the `hints` argument, but different operating systems implement different defaults in that case.

The final parameter to `getaddrinfo()`, `res`, is a pointer to a pointer to `struct addrinfo` and returns the address(es) found by `getaddrinfo()`.

If the call to `getaddrinfo()` succeeds, then its return value is 0. In this case, you should call `freeaddrinfo()` on `*res` when you've finished using the address. Here is an example of using `getaddrinfo()` to find the address(es) for `example.com`:

```
struct addrinfo hints;
memset(&hints, 0, sizeof(hints));
hints.ai_flags = AI_ALL;
struct addrinfo *peer_address;
if (getaddrinfo("example.com", 0, &hints, &peer_address)) {
    fprintf(stderr, "getaddrinfo() failed. (%d)\n", GETSOCKETERRNO());
    return 1;
}
```

Note that we first zero out `hints` using a call to `memset()`. We then set the `AI_ALL` flag, which specifies that we want both IPv4 and IPv6 addresses returned. This even returns addresses that we don't have a network adapter for. If you only want addresses that your machine can practically connect to, then use `AI_ALL | AI_ADDRCONFIG` for the `ai_flags` field. We can leave the other fields of `hints` as their defaults.

We then declare a pointer to hold the return address list: `struct addrinfo *peer_address`.

If the call to `getaddrinfo()` succeeds, then `peer_address` holds the first address result. The next result, if any, is in `peer_address->ai_next`.

We can loop through all the returned addresses with the following code:

```
struct addrinfo *address = peer_address;
do {
    /* Work with address... */
} while ((address = address->ai_next));
```

When we've finished using `peer_address`, we should free it with the following code:

```
freeaddrinfo(peer_address);
```

Now that we can convert a text address or name into an `addrinfo` structure, it is useful to see how to convert the `addrinfo` structure back into a text format. Let's look at that now.

Using getnameinfo()

`getnameinfo()` can be used to convert an `addrinfo` structure back into a text format. It works with both IPv4 and IPv6. It also, optionally, converts a port number into a text format number or service name.

The declaration for `getnameinfo()` can be seen in the following code:

```
int getnameinfo(const struct sockaddr *addr, socklen_t addrlen,
        char *host, socklen_t hostlen,
        char *serv, socklen_t servlen, int flags);
```

The first two parameters are passed in from the `ai_addr` and `ai_addrlen` fields of `struct addrinfo`.

The next two parameters, `host` and `hostlen`, specify a character buffer and buffer length to store the hostname or IP address text.

The following two parameters, `serv` and `servlen`, specify the buffer and length to store the service name.

If you don't need both the hostname and the service, you can optionally pass in only one of either `host` or `serv`.

Flags can be a bitwise OR combination of the following:

- `NI_NAMEREQD` requires `getnameinfo()` to return a hostname and not an address. By default, `getnameinfo()` tries to return a hostname but returns an address if it can't. `NI_NAMEREQD` will cause an error to be returned if the hostname cannot be determined.
- `NI_DGRAM` specifies that the service is UDP-based rather than TCP-based. This is only important for ports that have different standard services for UDP versus TCP. This flag is ignored if `NI_NUMERICSERV` is set.
- `NI_NUMERICHOST` requests that `getnameinfo()` returns the IP address and not a hostname.
- `NI_NUMERICSERV` requests that `getnameinfo()` returns the port number and not a service name.

For example, we can use `getnameinfo()` as follows:

```
char host[100];
char serv[100];
getnameinfo(address->ai_addr, address->ai_addrlen,
        host, sizeof(host),
        serv, sizeof(serv),
        0);

printf("%s %s\n", host, serv);
```

In the preceding code, `getnameinfo()` attempts to perform a reverse DNS lookup. This works like the DNS queries we've done in this chapter so far, but backward. A DNS query asks *which IP address does this hostname point to?* A reverse DNS query asks instead, *which hostname does this IP address point to?* Keep in mind that this is not a one-to-one relationship. Many hostnames can point to one IP address, but an IP address can store a DNS record for only one hostname. In fact, reverse DNS records are not even set for many IP addresses.

If `address` is a `struct addrinfo` with the address for `example.com` port 80 (`http`), then the preceding code might print as follows:

```
example.com http
```

If the code prints something different for you, it's probably working as intended. It is dependent on which address is in `address` and how the reverse DNS is set up for that IP address. Try it with a different address.

If, instead of the hostname, we would like the IP address, we can modify our code to the following:

```
char host[100];
char serv[100];
getnameinfo(address->ai_addr, address->ai_addrlen,
        host, sizeof(host),
        serv, sizeof(serv),
        NI_NUMERICHOST | NI_NUMERICSERV);

printf("%s %s\n", host, serv);
```

In the case of the previous code, it might print the following:

```
93.184.216.34 80
```

Using `NI_NUMERICHOST` generally runs much faster too, as it doesn't require `getnameinfo()` to send off any reverse DNS queries.

Alternative functions

Two widely-available functions that replicate `getaddrinfo()` are `gethostbyname()` and `getservbyname()`. The `gethostbyname()` function is obsolete and has been removed from the newer POSIX standards. Furthermore, I recommend against using these functions in new code, because they introduce an IPv4 dependency. It's very possible to use `getaddrinfo()` in such a way that your program does not need to be aware of IPv4 versus IPv6, but still supports both.

IP lookup example program

To demonstrate the getaddrinfo() and getnameinfo() functions, we will implement a short program. This program takes a name or IP address for its only argument. It then uses getaddrinfo() to resolve that name or that IP address into an address structure, and the program prints that IP address using getnameinfo() for the text conversion. If multiple addresses are associated with a name, it prints each of them. It also indicates any errors.

To begin with, we need to include our required header for this chapter. We also define AI_ALL for systems that are missing it. The code for this is as follows:

```
/*lookup.c*/

#include "chap05.h"

#ifndef AI_ALL
#define AI_ALL 0x0100
#endi
```

We can then begin the main() function and check that the user passed in a hostname to lookup. If the user doesn't pass in a hostname, we print a helpful reminder. The code for this is as follows:

```
/*lookup.c continued*/

int main(int argc, char *argv[]) {

    if (argc < 2) {
        printf("Usage:\n\tlookup hostname\n");
        printf("Example:\n\tlookup example.com\n");
        exit(0);
    }
```

We need the following code to initialize Winsock on Windows platforms:

```
/*lookup.c continued*/

#if defined(_WIN32)
    WSADATA d;
    if (WSAStartup(MAKEWORD(2, 2), &d)) {
        fprintf(stderr, "Failed to initialize.\n");
        return 1;
    }
#endif
```

We can then call `getaddrinfo()` to convert the hostname or address into a `struct addrinfo`. The code for that is as follows:

```
/*lookup.c continued*/

    printf("Resolving hostname '%s'\n", argv[1]);
    struct addrinfo hints;
    memset(&hints, 0, sizeof(hints));
    hints.ai_flags = AI_ALL;
    struct addrinfo *peer_address;
    if (getaddrinfo(argv[1], 0, &hints, &peer_address)) {
        fprintf(stderr, "getaddrinfo() failed. (%d)\n", GETSOCKETERRNO());
        return 1;
    }
```

The previous code first prints the hostname or address that is passed in as the first command-line argument. This argument is stored in `argv[1]`. We then set `hints.ai_flags = AI_ALL` to specify that we want all available addresses of any type, including both IPv4 and IPv6 addresses.

`getaddrinfo()` is then called with `argv[1]`. We pass 0 in for the service argument because we don't care about the port number. We are only trying to resolve an address. If `argv[1]` contains a name, such as `example.com`, then our operating system performs a DNS query (assuming the hostname isn't already in the local cache). If `argv[1]` contains an address such as `192.168.1.1`, then `getaddrinfo()` simply fills in the resulting `struct addrinfo` as needed.

If the user passed in an invalid address or a hostname for which no record could be found, then `getaddrinfo()` returns a non-zero value. In that case, our previous code prints out the error.

Now that `peer_address` holds the desired address(es), we can use `getnameinfo()` to convert them to text. The following code does that:

```
/*lookup.c continued*/

    printf("Remote address is:\n");
    struct addrinfo *address = peer_address;
    do {
        char address_buffer[100];
        getnameinfo(address->ai_addr, address->ai_addrlen,
                address_buffer, sizeof(address_buffer),
                0, 0,
                NI_NUMERICHOST);
        printf("\t%s\n", address_buffer);
    } while ((address = address->ai_next));
```

This code works by first storing `peer_address` in a new variable, `address`. We then enter a loop. `address_buffer[]` is declared to store the text address, and we call `getnameinfo()` to fill in that address. The last parameter to `getnameinfo()`, `NI_NUMERICHOST`, indicates that we want it to put the IP address into `address_buffer` and not a hostname. The address buffer can then simply be printed out with `printf()`.

If `getaddrinfo()` returned multiple addresses, then the next address is pointed to by `address->ai_next`. We assign `address->ai_next` to `address` and loop if it is non-zero. This is a simple example of walking through a linked list.

After we've printed our address, we should use `freeaddrinfo()` to free the memory allocated by `getaddrinfo()`. We should also call the Winsock cleanup function on Windows. We can do both with the following code:

```
/*lookup.c continued*/

    freeaddrinfo(peer_address);

#if defined(_WIN32)
    WSACleanup();
#endif

    return 0;
}
```

That concludes our `lookup` program.

You can compile and run `lookup.c` on Linux and macOS by using the following command:

```
gcc lookup.c -o lookup
./lookup example.com
```

Compiling and running on Windows with MinGW is done in the following way:

```
gcc lookup.c -o lookup.exe -lws2_32
lookup.exe example.com
```

The following screenshot is an example of using `lookup` to print the IP addresses for `example.com`:

```
Command Prompt                              —    □    ×

$ gcc lookup.c -o lookup.exe -lws2_32

$ lookup example.com
Resolving hostname 'example.com'
Remote address is:
        93.184.216.34
        2606:2800:220:1:248:1893:25c8:1946

$
```

Although `getaddrinfo()` makes performing DNS lookups easy, it is useful to know what happens behind the scenes. We will now look at the DNS protocol in more detail.

The DNS protocol

When a client wants to resolve a hostname into an IP address, it sends a DNS query to a DNS server. This is typically done over UDP using port 53. The DNS server then performs the lookup, if possible, and returns an answer. The following diagram illustrates this transaction:

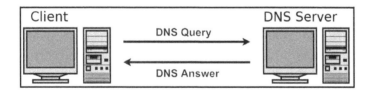

If the query (or, more commonly, the answer) is too large to fit into one UDP packet, then the query can be performed over TCP instead of UDP. In this case, the size of the query is sent over TCP as a 16-bit value, and then the query itself is sent. This is called **TCP fallback** or **DNS transport over TCP**. However, UDP works for most cases, and UDP is how DNS is used the vast majority of the time.

It's also important to note that the client must know the IP address of at least one DNS server. If the client doesn't know of any DNS servers, then it has a sort of chicken-and-egg problem. DNS servers are usually provided by your ISP.

The actual UDP data format is simple and follows the same basic format for both the query and the answer.

DNS message format

The following illustration describes the DNS message format:

DNS Message Format	
Header	Information about the message
Question	Question for the name server
Answer	Answer(s) to the question
Authority	Pointers to other name servers
Additional	Additional information

Every DNS message follows that format, although a query would leave the **Answer**, **Authority**, and **Additional** sections blank. A DNS response commonly doesn't use **Authority** or **Additional**. We won't concern ourselves with the **Authority** or **Additional** sections, as they are not needed for typical DNS queries.

DNS message header format

The header is exactly 12 bytes long and is exactly the same for a DNS query or DNS response. The **Header Format** is illustrated graphically in the following diagram:

```
                        Header Format
        0  1  2  3  4  5  6  7  8  9  10 11 12 13 14 15
       ┌─────────────────────────────────────────────┐
       │                      ID                       │
       ├───┬──────────┬──┬──┬──┬──┬───────┬────────────┤
       │QR │  Opcode  │AA│TC│RD│RA│   Z   │   RCODE    │
       ├───┴──────────┴──┴──┴──┴──┴───────┴────────────┤
       │                   QDCOUNT                      │
       ├───────────────────────────────────────────────┤
       │                   ANCOUNT                      │
       ├───────────────────────────────────────────────┤
       │                   NSCOUNT                      │
       ├───────────────────────────────────────────────┤
       │                   ARCOUNT                      │
       └───────────────────────────────────────────────┘
```

As the preceding diagram illustrates, the DNS message header contains 13 fields—**ID**, **QR**, **OPCODE**, **AA**, **TC**, **RD**, **RA**, **Z**, **RCODE**, **QDCOUNT**, **ANCOUNT**, **NSCOUNT**, and **ARCOUNT**:

- **ID** is any 16-bit value that is used to identify the DNS message. The client is allowed to put any 16 bits into the DNS query, and the DNS server copies those same 16 bits into the DNS response **ID**. This is useful to allow the client to match up which response is in reply to which query, in cases where the client is sending multiple queries.

- **QR** is a 1-bit field. It is set to 0 to indicate a DNS query or 1 to indicate a DNS response.

- **Opcode** is a 4-bit field, which specifies the type of query. 0 indicates a standard query. 1 indicates a reverse query to resolve an IP address into a name. 2 indicates a server status request. Other values (3 through 15) are reserved.

- **AA** indicates an authoritative answer.

- **TC** indicates that the message was truncated. In this case, it should be re-sent using TCP.

- **RD** should be set if recursion is desired. We leave this bit set to indicate that we want the DNS server to contact additional servers until it can complete our request.

- **RA** indicates in a response whether the DNS server supports recursion.

- **Z** is unused and should be set to 0.

- **RCODE** is set in a DNS response to indicate the error condition. Its possible values are as follows:

RCODE	Description
0	No error
1	Format error
2	Server failure
3	Name error
4	Not implemented
5	Refused

Please see **RFC 1035**: *DOMAIN NAMES – IMPLEMENTATION AND SPECIFICATION*, in the *Further reading* section of this chapter for more information on the meaning of these values.

- **QDCOUNT, ANCOUNT, NSCOUNT**, and **ARCOUNT** indicate the number of records in their corresponding sections. **QDCOUNT** indicates the number of questions in a DNS query. It is interesting that **QDCOUNT** is a 16-bit value capable of storing large numbers, and yet no real-world DNS server allows more than one question per message. **ANCOUNT** indicates the number of answers, and it is common for a DNS server to return multiple answers in one message.

Question format

The DNS **Question Format** consists of a name followed by two 16-bit values—QTYPE and QCLASS. This **Question Format** is illustrated as follows:

```
                    Question Format
     0  1  2  3  4  5  6  7  8  9  10 11 12 13 14 15

                        NAME

                        QTYPE
                        QCLASS
```

QTYPE indicates the record type we are asking for, and **QCLASS** should be set to 1 to indicate the internet.

The name field involves a special encoding. First, a hostname should be broken up into its individual labels. For example, www.example.com would be broken up into www, example, and com. Then, each label should be prepended with 1 byte, indicating the label length. Finally, the entire name ends with a 0 byte.

Visually, the name www.example.com is encoded as follows:

Name Example															
0	1	2	3	4	5	6	7	8	9	10	11	12	13	14	15
3								'w'							
'w'								'w'							
7								'e'							
'x'								'a'							
'm'								'p'							
'l'								'e'							
3								'c'							
'o'								'm'							
0															

If the **QTYPE** and **QCLASS** fields were appended to the preceding name example, then it could make up an entire DNS question.

A DNS response is sometimes required to repeat the same name multiple times. In this case, a DNS server may encode a pointer to an earlier name instead of sending the same name multiple times. A pointer is indicated by a 16-bit value with the two most significant bits set. The lower 14 bits indicate the pointer value. This 14-bit value specifies the location of the name as an offset from the beginning of the message. Having the two most significant bits reserved has an additional side-effect of limiting labels to 63 characters. A longer name would require setting both high bits in the label length specifier, but if both high bits are set, it indicates a pointer and not a label length!

The answer format is similar to the question format but with a few more fields. Let's look at that next.

Answer format

The DNS answer format consists of the same three fields that questions have; namely, a name followed by a 16-bit TYPE and a 16-bit CLASS. The answer format then has a 32-bit TTL field. This field specifies how many seconds the answer is allowed to be cached for. TTL is followed by a 16-bit length specifier, RDLENGTH, followed by data. The data is RDLENGTH long, and the data's interpretation is dependent upon the type specified by TYPE.

Visually, the answer format is shown in the following diagram:

```
                        Answer Format
          0  1  2  3  4  5  6  7  8  9  10 11 12 13 14 15
        ┌──────────────────────────────────────────────┐
        │                    NAME                        │
        ├──────────────────────────────────────────────┤
        │                    TYPE                        │
        ├──────────────────────────────────────────────┤
        │                    CLASS                       │
        ├──────────────────────────────────────────────┤
        │                     TTL                        │
        ├──────────────────────────────────────────────┤
        │                  RDLENGTH                      │
        ├──────────────────────────────────────────────┤
        │                    RDATA                       │
        │                                                │
        └──────────────────────────────────────────────┘
```

Keep in mind that most DNS servers use name pointers in their answer names. This is because the DNS response will have already included the relevant name in the question section. The answer can simply point back to that, rather than encoding the entire name a second (or third) time.

Whenever sending binary data over the network, the issue of byte order becomes relevant. Let's consider this now.

Endianness

The term **endianness** refers to the order in which individual bytes are stored in memory or sent over a network.

Whenever we read a multi-byte number from a DNS message, we should be aware that it's in big-endian format (or so-called network byte order). The computer you're using likely uses little-endian format, although we are careful to write our code in an endian-independent manner throughout this book. We accomplish this by avoiding the conversion of multiple bytes directly to integers, and instead we interpret bytes one at a time.

For example, consider a message with a single 8-bit value, such as 0x05. We know that the value of that message is 5. Bytes are sent atomically over a network link, so we also know that anyone receiving our message can unambiguously interpret that message as 5.

The issue of endianness comes into play when we need more than one byte to store our number. Imagine that we want to send the number 999. This number is too big to fit into 1 byte, so we have to break it up into 2 bytes—a 16-bit value. Because $999 = (3 * 2^8) + 231$, we know that the high-order byte stores 3 while the low-order byte stores 231. In hexadecimal, the number 999 is 0x03E7. The question is whether to send the high-order or the low-order byte over the network first.

Network byte order, which is used by the DNS protocol, specifies that the high-order byte is sent first. Therefore, the number 999 is sent over the network as a 0x03 followed by 0xE7.

See the *Further reading* section of this chapter for more information.

Let's now look at encoding an entire DNS query.

A simple DNS query

To perform a simple DNS query, we would put an arbitrary number into ID, set the RD bit to 1, and set QDCOUNT to 1. We would then add a question after the header. That data would be sent as a UDP packet to port 53 of a DNS server.

A hand-constructed DNS for `example.com` in C is as follows:

```
char dns_query[] = {0xAB, 0xCD,                            /* ID */
                    0x01, 0x00,                            /* Recursion */
                    0x00, 0x01,                            /* QDCOUNT */
                    0x00, 0x00,                            /* ANCOUNT */
                    0x00, 0x00,                            /* NSCOUNT */
                    0x00, 0x00,                            /* ARCOUNT */
                    7, 'e', 'x', 'a', 'm', 'p', 'l', 'e',  /* label */
                    3, 'c', 'o', 'm',                      /* label */
                    0,                                     /* End of name */
                    0x00, 0x01,                            /* QTYPE = A */
                    0x00, 0x01                             /* QCLASS */
                    };
```

This data could be sent as is to a DNS server over port 53.

The DNS server, if successful, sends a UDP packet back as a response. This packet has ID set to match our query. QR is set to indicate a response. QDCOUNT is set to 1, and our original question is included. ANCOUNT is some small positive integer that indicates the number of answers included in the message.

In the next section, we'll implement a program to send and receive DNS messages.

A DNS query program

We will now implement a utility to send DNS queries to a DNS server and receive the DNS response.

This should not normally be needed in the field. It is, however, a good opportunity to better understand the DNS protocol and to get experience of sending binary UDP packets.

We begin with a function to print a name from a DNS message.

Printing a DNS message name

DNS encodes names in a particular way. Normally, each label is indicated by its length, followed by its text. A number of labels can be repeated, and then the name is terminated with a single 0 byte.

If a length has its two highest bits set (that is, 0xc0), then it and the next byte should be interpreted as a pointer instead.

We must also be aware at all times that the DNS response from the DNS server could be ill-formed or corrupted. We must try to write our program in such a way that it won't crash if it receives a bad message. This is easier said than done.

The declaration for our name-printing function looks like this:

```
/*dns_query.c*/

const unsigned char *print_name(const unsigned char *msg,
        const unsigned char *p, const unsigned char *end);
```

We take `msg` to be a pointer to the message's beginning, `p` to be a pointer to the name to print, and `end` to be a pointer to one past the end of the message. `end` is required so that we can check that we're not reading past the end of the received message. `msg` is required for the same reason, but also so that we can interpret name pointers.

Inside the `print_name` function, our code checks that a proper name is even possible. Because a name should consist of at least a length and some text, we can return an error if `p` is already within two characters of the end. The code for that check is as follows:

```
/*dns_query.c*/

    if (p + 2 > end) {
        fprintf(stderr, "End of message.\n"); exit(1);}
```

We then check to see if `p` points to a name pointer. If it does, we interpret the pointer and call `print_name` recursively to print the name that is pointed to. The code for this is as follows:

```
/*dns_query.c*/

    if ((*p & 0xC0) == 0xC0) {
        const int k = ((*p & 0x3F) << 8) + p[1];
        p += 2;
        printf(" (pointer %d) ", k);
        print_name(msg, msg+k, end);
        return p;
```

Note that `0xC0` in binary is `0b11000000`. We use `(*p & 0xC0) == 0xC0` to check for a name pointer. In that case, we take the lower 6 bits of `*p` and all 8 bits of `p[1]` to indicate the pointer. We know that `p[1]` is still within the message because of our earlier check that `p` was at least 2 bytes from the end. Knowing the name pointer, we can pass a new value of `p` to `print_name()`.

If the name is not a pointer, we simply print it one label at a time. The code for printing the name is as follows:

```
/*dns_query.c*/

    } else {
        const int len = *p++;
        if (p + len + 1 > end) {
            fprintf(stderr, "End of message.\n"); exit(1);}

        printf("%.*s", len, p);
        p += len;
        if (*p) {
            printf(".");
            return print_name(msg, p, end);
        } else {
            return p+1;
        }
    }
}
```

In the preceding code, *p is read into `len` to store the length of the current label. We are careful to check that reading `len + 1` bytes doesn't put us past the end of the buffer. We can then print the next `len` characters to the console. If the next byte isn't 0, then the name is continued, and we should print a dot to separate the labels. We call `print_name()` recursively to print the next label of the name.

If the next byte is 0, then that means the name is finished and we return.

That concludes the `print_name()` function.

Now we will devise a function that prints an entire DNS message.

Printing a DNS message

Using `print_name()` that we have just defined, we can now construct a function to print an entire DNS message to the screen. DNS messages share the same format for both the request and the response, so our function is able to print either.

The declaration for our function is as follows:

```
/*dns_query.c*/

void print_dns_message(const char *message, int msg_length);
```

`print_dns_message()` takes a pointer to the start of the message, and an `int` data type indicates the message's length.

Inside `print_dns_message()`, we first check that the message is long enough to be a valid DNS message. Recall that the DNS header is 12 bytes long. If a DNS message is less than 12 bytes, we can easily reject it as an invalid message. This also ensures that we can read at least the header without worrying about reading past the end of the received data.

The code for checking the DNS message length is as follows:

```
/*dns_query.c*/

if (msg_length < 12) {
    fprintf(stderr, "Message is too short to be valid.\n");
    exit(1);
}
```

We then copy the `message` pointer into a new variable, `msg`. We define `msg` as an `unsigned char` pointer, which makes certain calculations easier to work with.

```
/*dns_query.c*/

const unsigned char *msg = (const unsigned char *)message;
```

If you want to print out the entire raw DNS message, you can do that with the following code:

```
/*dns_query.c*/

int i;
for (i = 0; i < msg_length; ++i) {
    unsigned char r = msg[i];
    printf("%02d:   %02X  %03d  '%c'\n", i, r, r, r);
}
printf("\n");
```

Be aware that running the preceding code will print out many lines. This can be annoying, so I would recommend using it only if you are curious about seeing the raw DNS message.

The message ID can be printed very easily. Recall that the message ID is simply the first two bytes of the message. The following code prints it in a nice hexadecimal format:

```
/*dns_query.c*/

printf("ID = %0X %0X\n", msg[0], msg[1]);
```

Next, we get the QR bit from the message header. This bit is the most significant bit of `msg[2]`. We use the bitmask `0x80` to see whether it is set. If it is, we know that the message is a response; otherwise, it's a query. The following code reads QR and prints a corresponding message:

```
/*dns_query.c*/

    const int qr = (msg[2] & 0x80) >> 7;
    printf("QR = %d %s\n", qr, qr ? "response" : "query");
```

The OPCODE, AA, TC, and RD fields are read in much the same way as QR. The code for printing them is as follows:

```
/*dns_query.c*/

    const int opcode = (msg[2] & 0x78) >> 3;
    printf("OPCODE = %d ", opcode);
    switch(opcode) {
        case 0: printf("standard\n"); break;
        case 1: printf("reverse\n"); break;
        case 2: printf("status\n"); break;
        default: printf("?\n"); break;
    }

    const int aa = (msg[2] & 0x04) >> 2;
    printf("AA = %d %s\n", aa, aa ? "authoritative" : "");

    const int tc = (msg[2] & 0x02) >> 1;
    printf("TC = %d %s\n", tc, tc ? "message truncated" : "");

    const int rd = (msg[2] & 0x01);
    printf("RD = %d %s\n", rd, rd ? "recursion desired" : "");
```

Finally, we can read in RCODE for response-type messages. Since RCODE can have several different values, we use a `switch` statement to print them. Here is the code for that:

```
/*dns_query.c*/

    if (qr) {
        const int rcode = msg[3] & 0x07;
        printf("RCODE = %d ", rcode);
        switch(rcode) {
            case 0: printf("success\n"); break;
            case 1: printf("format error\n"); break;
            case 2: printf("server failure\n"); break;
            case 3: printf("name error\n"); break;
            case 4: printf("not implemented\n"); break;
```

```
            case 5: printf("refused\n"); break;
            default: printf("?\n"); break;
        }
        if (rcode != 0) return;
    }
```

The next four fields in the header are the question count, the answer count, the name server count, and the additional count. We can read and print them in the following code:

```
/*dns_query.c*/

    const int qdcount = (msg[4] << 8) + msg[5];
    const int ancount = (msg[6] << 8) + msg[7];
    const int nscount = (msg[8] << 8) + msg[9];
    const int arcount = (msg[10] << 8) + msg[11];

    printf("QDCOUNT = %d\n", qdcount);
    printf("ANCOUNT = %d\n", ancount);
    printf("NSCOUNT = %d\n", nscount);
    printf("ARCOUNT = %d\n", arcount);
```

That concludes reading the DNS message header (the first 12 bytes).

Before reading the rest of the message, we define two new variables as follows:

```
/*dns_query.c*/

    const unsigned char *p = msg + 12;
    const unsigned char *end = msg + msg_length;
```

In the preceding code, the p variable is used to walk through the message. We set the end variable to one past the end of the message. This is to help us detect whether we're about to read past the end of the message – a situation we certainly wish to avoid!

We read and print each question in the DNS message with the following code:

```
/*dns_query.c*/

    if (qdcount) {
        int i;
        for (i = 0; i < qdcount; ++i) {
            if (p >= end) {
                fprintf(stderr, "End of message.\n"); exit(1);}

            printf("Query %2d\n", i + 1);
            printf("  name: ");

            p = print_name(msg, p, end); printf("\n");
```

```
        if (p + 4 > end) {
            fprintf(stderr, "End of message.\n"); exit(1);}

        const int type = (p[0] << 8) + p[1];
        printf("  type: %d\n", type);
        p += 2;

        const int qclass = (p[0] << 8) + p[1];
        printf("  class: %d\n", qclass);
        p += 2;
    }
}
```

Although no real-world DNS server will accept a message with multiple questions, the DNS RFC does clearly define the format to encode multiple questions. For that reason, we make our code loop through each question using a `for` loop. First, the `print_name()` function, which we defined earlier, is called to print the question name. We then read in and print out the question type and class.

Printing the answer, authority, and additional sections is slightly more difficult than the question section. These sections start the same way as the question section – with a name, a type, and a class. The code for reading the name, type, and class is as follows:

```
/*dns_query.c*/

    if (ancount || nscount || arcount) {
        int i;
        for (i = 0; i < ancount + nscount + arcount; ++i) {
            if (p >= end) {
                fprintf(stderr, "End of message.\n"); exit(1);}

            printf("Answer %2d\n", i + 1);
            printf("  name: ");

            p = print_name(msg, p, end); printf("\n");

            if (p + 10 > end) {
                fprintf(stderr, "End of message.\n"); exit(1);}

            const int type = (p[0] << 8) + p[1];
            printf("  type: %d\n", type);
            p += 2;

            const int qclass = (p[0] << 8) + p[1];
            printf("  class: %d\n", qclass);
            p += 2;
```

Note that, in the preceding code, we stored the class in a variable called `qclass`. This is to be nice to our C++ friends, who are not allowed to use `class` as a variable name.

We then expect to find a 16-bit TTL field, and a 16-bit data length field. The TTL field tells us how many seconds we are allowed to cache an answer for. The data length field tells us how many bytes of additional data are included for the answer. We read TTL and the data length in the following code:

```
/*dns_query.c*/

        const unsigned int ttl = (p[0] << 24) + (p[1] << 16) +
            (p[2] << 8) + p[3];
        printf("   ttl: %u\n", ttl);
        p += 4;

        const int rdlen = (p[0] << 8) + p[1];
        printf(" rdlen: %d\n", rdlen);
        p += 2;
```

Before we can read in the data of the `rdlen` length, we should check that we won't read past the end of the message. The following code achieves that:

```
/*dns_query.c*/

        if (p + rdlen > end) {
            fprintf(stderr, "End of message.\n"); exit(1);}
```

We can then try to interpret the answer data. Each record type stores different data. We need to write code to display each type. For our purposes, we limit this to the A, MX, AAAA, TXT, and CNAME records. The code to print each type is as follows:

```
/*dns_query.c*/

        if (rdlen == 4 && type == 1) { /* A Record */
            printf("Address ");
            printf("%d.%d.%d.%d\n", p[0], p[1], p[2], p[3]);

        } else if (type == 15 && rdlen > 3) { /* MX Record */
            const int preference = (p[0] << 8) + p[1];
            printf("  pref: %d\n", preference);
            printf("MX: ");
            print_name(msg, p+2, end); printf("\n");

        } else if (rdlen == 16 && type == 28) { /* AAAA Record */
            printf("Address ");
            int j;
            for (j = 0; j < rdlen; j+=2) {
```

```
            printf("%02x%02x", p[j], p[j+1]);
            if (j + 2 < rdlen) printf(":");
        }
        printf("\n");

    } else if (type == 16) { /* TXT Record */
        printf("TXT: '%.*s'\n", rdlen-1, p+1);

    } else if (type == 5) { /* CNAME Record */
        printf("CNAME: ");
        print_name(msg, p, end); printf("\n");
    }
    p += rdlen;
```

We can then finish the loop. We check that all the data was read and print a message if it wasn't. If our program is correct, and if the DNS message is properly formatted, we should have read all the data with nothing left over. The following code checks this:

```
/*dns_query.c*/

        }
    }

    if (p != end) {
        printf("There is some unread data left over.\n");
    }

    printf("\n");
```

That concludes the `print_dns_message()` function.

We can now define our `main()` function to create the DNS query, send it to a DNS server, and await a response.

Sending the query

We start `main()` with the following code:

```
/*dns_query.c*/

int main(int argc, char *argv[]) {

    if (argc < 3) {
        printf("Usage:\n\tdns_query hostname type\n");
        printf("Example:\n\tdns_query example.com aaaa\n");
        exit(0);
```

```
    }

    if (strlen(argv[1]) > 255) {
        fprintf(stderr, "Hostname too long.");
        exit(1);
    }
```

The preceding code checks that the user passed in a hostname and record type to query. If they didn't, it prints a helpful message. It also checks that the hostname isn't more than 255 characters long. Hostnames longer than that aren't allowed by the DNS standard, and checking it now ensures that we don't need to allocate too much memory.

We then try to interpret the record type requested by the user. We support the following options – a, aaaa, txt, mx, and any. The code to read in those types and store their corresponding DNS integer value is as follows:

```
/*dns_query.c*/

    unsigned char type;
    if (strcmp(argv[2], "a") == 0) {
        type = 1;
    } else if (strcmp(argv[2], "mx") == 0) {
        type = 15;
    } else if (strcmp(argv[2], "txt") == 0) {
        type = 16;
    } else if (strcmp(argv[2], "aaaa") == 0) {
        type = 28;
    } else if (strcmp(argv[2], "any") == 0) {
        type = 255;
    } else {
        fprintf(stderr, "Unknown type '%s'. Use a, aaaa, txt, mx, or any.",
                argv[2]);
        exit(1);
    }
```

Like all of our previous programs, we need to initialize Winsock. The code for that is as follows:

```
/*dns_query.c*/

#if defined(_WIN32)
    WSADATA d;
    if (WSAStartup(MAKEWORD(2, 2), &d)) {
        fprintf(stderr, "Failed to initialize.\n");
        return 1;
    }
#endif
```

Our program connects to 8.8.8.8, which is a public DNS server run by Google. Refer to Chapter 1, *An Introduction to Networks and Protocols, Domain Names*, for a list of additional public DNS servers you can use.

Recall that we are connecting on UDP port 53. We use getaddrinfo() to set up the required structures for our socket with the following code:

```
/*dns_query.c*/

    printf("Configuring remote address...\n");
    struct addrinfo hints;
    memset(&hints, 0, sizeof(hints));
    hints.ai_socktype = SOCK_DGRAM;
    struct addrinfo *peer_address;
    if (getaddrinfo("8.8.8.8", "53", &hints, &peer_address)) {
        fprintf(stderr, "getaddrinfo() failed. (%d)\n", GETSOCKETERRNO());
        return 1;
    }
```

We then create our socket using the data returned from getaddrinfo(). The following code does that:

```
/*dns_query.c*/

    printf("Creating socket...\n");
    SOCKET socket_peer;
    socket_peer = socket(peer_address->ai_family,
            peer_address->ai_socktype, peer_address->ai_protocol);
    if (!ISVALIDSOCKET(socket_peer)) {
        fprintf(stderr, "socket() failed. (%d)\n", GETSOCKETERRNO());
        return 1;
    }
```

Our program then constructs the data for the DNS query message. The first 12 bytes compose the header and are known at compile time. We can store them with the following code:

```
/*dns_query.c*/

    char query[1024] = {0xAB, 0xCD, /* ID */
                        0x01, 0x00, /* Set recursion */
                        0x00, 0x01, /* QDCOUNT */
                        0x00, 0x00, /* ANCOUNT */
                        0x00, 0x00, /* NSCOUNT */
                        0x00, 0x00 /* ARCOUNT */};
```

The preceding code sets our query's ID to 0xABCD, sets a recursion request, and indicates that we are attaching 1 question. As mentioned earlier, 1 is the only number of questions supported by real-world DNS servers.

We then need to encode the user's desired hostname into the query. The following code does that:

```
/*dns_query.c*/

    char *p = query + 12;
    char *h = argv[1];

    while(*h) {
        char *len = p;
        p++;
        if (h != argv[1]) ++h;

        while(*h && *h != '.') *p++ = *h++;
        *len = p - len - 1;
    }

    *p++ = 0;
```

The preceding code first sets a new pointer, p, to the end of the query header. We will be adding to the query starting at p. We also define a pointer, h, which we use to loop through the hostname.

We can loop while *h != 0 because *h is equal to zero when we've finished reading the hostname. Inside the loop, we use the len variable to store the position of the label beginning. The value in this position needs to be set to indicate the length of the upcoming label. We then copy characters from *h to *p until we find a dot or the end of the hostname. If either is found, the code sets *len equal to the label length. The code then loops into the next label.

Finally, outside the loop, we add a terminating 0 byte to finish the name section of the question.

We then add the question type and question class to the query with the following code:

```
/*dns_query.c*/

    *p++ = 0x00; *p++ = type; /* QTYPE */
    *p++ = 0x00; *p++ = 0x01; /* QCLASS */
```

We can then calculate the query size by comparing `p` to the query beginning. The code for figuring the total query size is as follows:

```
/*dns_query.c*/

    const int query_size = p - query;
```

Now, with the query message formed, and its length known, we can use `sendto()` to transmit the DNS query to the DNS server. The code for sending the query is as follows:

```
/*dns_query.c*/

    int bytes_sent = sendto(socket_peer,
            query, query_size,
            0,
            peer_address->ai_addr, peer_address->ai_addrlen);
    printf("Sent %d bytes.\n", bytes_sent);
```

For debugging purposes, we can also display the query we sent with the following code:

```
/*dns_query.c*/

    print_dns_message(query, query_size);
```

The preceding code is useful to see whether we've made any mistakes in encoding our query.

Now that the query has been sent, we await a DNS response message using `recvfrom()`. In a practical program, you may want to use `select()` here to time out. It could also be wise to listen for additional messages in the case that an invalid message is received first.

The code to receive and display the DNS response is as follows:

```
/*dns_query.c*/

    char read[1024];
    int bytes_received = recvfrom(socket_peer,
            read, 1024, 0, 0, 0);

    printf("Received %d bytes.\n", bytes_received);

    print_dns_message(read, bytes_received);
    printf("\n");
```

We can finish our program by freeing the address(es) from `getaddrinfo()` and cleaning up Winsock. The code to complete the `main()` function is as follows:

```
/*dns_query.c*/

    freeaddrinfo(peer_address);
    CLOSESOCKET(socket_peer);

#if defined(_WIN32)
    WSACleanup();
#endif

    return 0;
}
```

That concludes the `dns_query` program.

You can compile and run `dns_query.c` on Linux and macOS by running the following command:

```
gcc dns_query.c -o dns_query
./dns_query example.com a
```

Compiling and running on Windows with MinGW is done by using the following command:

```
gcc dns_query.c -o dns_query.exe -lws2_32
dns_query.exe example.com a
```

Try running `dns_query` with different domain names and different record types. In particular, try it with `mx` and `txt` records. If you're brave, try running it with the `any` record type. You may find the results interesting.

The following screenshot is an example of using `dns_query` to query the A record of `example.com`:

```
m1:Desktop honp$ gcc dns_query.c -o dns_query
m1:Desktop honp$ ./dns_query example.com a
Configuring remote address...
Creating socket...
Sent 29 bytes.
ID = AB CD
QR = 0 query
OPCODE = 0 standard
AA = 0
TC = 0
RD = 1 recursion desired
QDCOUNT = 1
ANCOUNT = 0
NSCOUNT = 0
ARCOUNT = 0
Query  1
  name: example.com
  type: 1
 class: 1

Received 45 bytes.
ID = AB CD
QR = 1 response
OPCODE = 0 standard
AA = 0
TC = 0
RD = 1 recursion desired
RCODE = 0 success
QDCOUNT = 1
ANCOUNT = 1
NSCOUNT = 0
ARCOUNT = 0
Query  1
  name: example.com
  type: 1
 class: 1
Answer  1
  name:  (pointer 12) example.com
  type: 1
 class: 1
   ttl: 20576
 rdlen: 4
Address 93.184.216.34

m1:Desktop honp$
```

The next screenshot shows `dns_query` querying the `mx` record of `gmail.com`:

```
● ○ ○                    🖳 Desktop — bash — 80×46         ◪
m1:Desktop honp$ ./dns_query gmail.com mx                 ▣
Configuring remote address...
Creating socket...
Sent 27 bytes.
ID = AB CD
QR = 0 query
OPCODE = 0 standard
AA = 0
TC = 0
RD = 1 recursion desired
QDCOUNT = 1
ANCOUNT = 0
NSCOUNT = 0
ARCOUNT = 0
Query  1
  name: gmail.com
  type: 15
 class: 1

Received 150 bytes.
ID = AB CD
QR = 1 response
OPCODE = 0 standard
AA = 0
TC = 0
RD = 1 recursion desired
RCODE = 0 success
QDCOUNT = 1
ANCOUNT = 5
NSCOUNT = 0
ARCOUNT = 0
Query  1
  name: gmail.com
  type: 15
 class: 1
Answer  1
  name:  (pointer 12) gmail.com
  type: 15
 class: 1
   ttl: 3599
 rdlen: 32
  pref: 30
MX: alt3.gmail-smtp-in.l.google. (pointer 18) com
Answer  2
  name:  (pointer 12) gmail.com
  type: 15
```

Note that UDP is not always reliable. If our DNS query is lost in transit, then `dns_query` hangs while waiting forever for a reply that never comes. This could be fixed by using the `select()` function to time out and retry.

Summary

This chapter was all about hostnames and DNS queries. We covered how the DNS works, and we learned that resolving a hostname can involve many UDP packets being sent over the network.

We looked at getaddrinfo() in more depth and showed why it is usually the preferred way to do a hostname lookup. We also looked at its sister function, getnameinfo(), which is capable of converting an address to text or even doing a reverse DNS query.

Finally, we implemented a program that sent DNS queries from scratch. This program was a good learning experience to better understand the DNS protocol, and it gave us a chance to gain experience in implementing a binary protocol. When implementing a binary protocol, we had to pay special attention to byte order. For the simple DNS message format, this was achieved by carefully interpreting bytes one at a time.

Now that we've worked with a binary protocol, DNS, we will move on to text-based protocols in the next few chapters. In the next chapter, we will learn about HTTP, the protocol used to request and retrieve web pages.

Questions

Try these questions to test your knowledge of this chapter:

1. Which function fills in an address needed for socket programming in a portable and protocol-independent way?
2. Which socket programming function can be used to convert an IP address back into a name?
3. A DNS query converts a name into an address, and a reverse DNS query converts an address back into a name. If you run a DNS query on a name, and then a reverse DNS query on the resulting address, do you always get back the name you started with?
4. What are the DNS record types used to return IPv4 and IPv6 addresses for a name?
5. Which DNS record type stores special information about email servers?
6. Does getaddrinfo() always return immediately? or can it block?
7. What happens when a DNS response is too large to fit into a single UDP packet?

The answers are in Appendix A, *Answers to Questions*.

Further reading

For more information about DNS, please refer to:

- **RFC 1034**: *DOMAIN NAMES – CONCEPTS AND FACILITIES* (`https://tools.ietf.org/html/rfc1034`)
- **RFC 1035**: *DOMAIN NAMES – IMPLEMENTATION AND SPECIFICATION* (`https://tools.ietf.org/html/rfc1035`)
- **RFC 3596**: *DNS Extensions to Support IP Version 6* (`https://tools.ietf.org/html/rfc3596`)

Section 2 - An Overview of Application Layer Protocols

2

In this section, the reader will learn about two of the most common higher-level protocols, HTTP and SMTP, which build on top of TCP.

The following chapters are in this section:

6
Building a Simple Web Client

Hypertext Transfer Protocol (**HTTP**) is the application protocol that powers the World Wide Web (WWW). Whenever you fire up your web browser to do an internet search, browse Wikipedia, or make a post on social media, you are using HTTP. Many mobile apps also use HTTP behind the scenes. It's safe to say that HTTP is one of the most widely used protocols on the internet.

In this chapter, we will look at the HTTP message format. We will then implement a C program, which can request and receive web pages.

The following topics are covered in this chapter:

- The HTTP message format
- HTTP request types
- Common HTTP headers
- HTTP response code
- HTTP message parsing
- Implementing an HTTP client
- Encoding form data (POST)
- HTTP file uploads

Technical requirements

The example programs from this chapter can be compiled with any modern C compiler. We recommend MinGW on Windows and GCC on Linux and macOS. See appendices B, C, and D for compiler setup.

The code for this book can be found at https://github.com/codeplea/Hands-On-Network-Programming-with-C.

From the command line, you can download the code for this chapter with the following command:

```
git clone https://github.com/codeplea/Hands-On-Network-Programming-with-C
cd Hands-On-Network-Programming-with-C/chap06
```

Each example program in this chapter runs on Windows, Linux, and macOS. When compiling on Windows, each example program requires linking with the Winsock library. This is accomplished by passing the -lws2_32 option to gcc.

We provide the exact commands needed to compile each example as they are introduced.

All of the example programs in this chapter require the same header files and C macros that we developed in Chapter 2, *Getting to Grips with Socket APIs*. For brevity, we put these statements in a separate header file, chap06.h, which we can include in each program. For an explanation of these statements, please refer to Chapter 2, *Getting to Grips with Socket APIs*.

The content of chap06.h is as follows:

```c
//chap06.h//

#if defined(_WIN32)
#ifndef _WIN32_WINNT
#define _WIN32_WINNT 0x0600
#endif
#include <winsock2.h>
#include <ws2tcpip.h>
#pragma comment(lib, "ws2_32.lib")

#else
#include <sys/types.h>
#include <sys/socket.h>
#include <netinet/in.h>
#include <arpa/inet.h>
#include <netdb.h>
#include <unistd.h>
#include <errno.h>

#endif


#if defined(_WIN32)
#define ISVALIDSOCKET(s) ((s) != INVALID_SOCKET)
#define CLOSESOCKET(s) closesocket(s)
#define GETSOCKETERRNO() (WSAGetLastError())
```

```
#else
#define ISVALIDSOCKET(s) ((s) >= 0)
#define CLOSESOCKET(s) close(s)
#define SOCKET int
#define GETSOCKETERRNO() (errno)
#endif


#include <stdio.h>
#include <stdlib.h>
#include <string.h>
#include <clock.h>
```

The HTTP protocol

HTTP is a text-based client-server protocol that runs over TCP. Plain HTTP runs over TCP port 80.

It should be noted that plain HTTP is mostly deprecated for security reasons. Today, sites should use HTTPS, the secure version of HTTP. HTTPS secures HTTP by merely running the HTTP protocol through a **Transport Layer Security** (**TLS**) layer. Therefore, everything we cover in this chapter regarding HTTP also applies to HTTPS. See Chapter 9, *Loading Secure Web Pages with HTTPS and OpenSSL*, for more information about HTTPS.

HTTP works by first having the web client send an HTTP request to the web server. Then, the web server responds with an HTTP response. Generally, the HTTP request indicates which resource the client is interested in, and the HTTP response delivers the requested resource.

Visually, the transaction is illustrated in the following graphic:

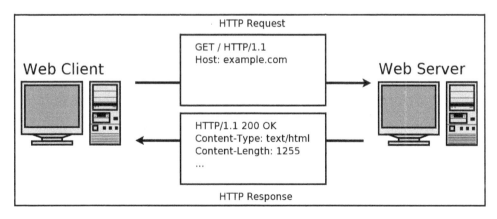

The preceding graphic illustrates a **GET** request. A **GET** request is used when the **Web Client** simply wants the **Web Server** to send it a document, image, file, web page, and so on. **GET** requests are the most common. They are what your browser sends to a **Web Server** while loading up a web page or downloading a file.

There are a few other request types that are also worth mentioning.

HTTP request types

Although GET requests are the most common, there are, perhaps, three request types that are commonly used. The three common HTTP request types are as follows:

- GET is used when the client wants to download a resource.
- HEAD is just like GET, except that the client only wants information about the resource instead of the resource itself. For example, if the client only wants to know the size of a hosted file, it could send a HEAD request.
- POST is used when the client needs to send information to the server. Your web browser typically uses a POST request when you submit an online form, for example. A POST request will typically cause a web server to change its state somehow. A web server could send an email, update a database, or change a file in response to a POST request.

In addition to GET, HEAD, and POST, there are a few more HTTP request types that are rarely used. They are as follows:

- PUT is used to send a document to the web server. PUT is not commonly used. POST is almost universally used to change the web server state.
- DELETE is used to request that the web server should delete a document or resource. Again, in practice, DELETE is rarely used. POST is commonly used to communicate web server updates of all types.
- TRACE is used to request diagnostic information from web proxies. Most web requests don't go through a proxy, and many web proxies don't fully support TRACE. Therefore, it's rare to need to use TRACE.
- CONNECT is sometimes used to initiate an HTTP connection through a proxy server.
- OPTIONS is used to ask which HTTP request types are supported by the server for a given resource. A typical web server that implements OPTIONS may respond with something similar to Allow: OPTIONS, GET, HEAD, POST. Many common web servers don't support OPTIONS.

If you send a request that the web server doesn't support, then the server should respond with a `400 Bad Request` code.

Now that we've seen the types of HTTP requests, let's look at the request format in more detail.

HTTP request format

If you open your web browser and navigate to `http://www.example.com/page1.htm`, your browser will need to send an HTTP request to the web server at `www.example.com`. That HTTP request may look like this:

```
GET /page1.htm HTTP/1.1
User-Agent: Mozilla/5.0 (Windows NT 10.0; Win64; x64) AppleWebKit/537.36
Accept-Language: en-US
Accept: text/html,application/xhtml+xml,application/xml;q=0.9,*/*;q=0.8
Accept-Encoding: gzip, deflate
Host: example.com
Connection: Keep-Alive
```

As you can see, the browser sends a `GET` request by default. This `GET` request is asking the server for the document `/page1.htm`. A `GET` request consists of HTTP headers only. There is no HTTP body because the client isn't sending data to the server. The client is only requesting data from the server. In contrast, a `POST` request would contain an HTTP body.

The first line of an HTTP request is called the **request line**. The request line consists of three parts – the request type, the document path, and the protocol version. Each part is separated by a space. In the preceding example, the request line is `GET /page1.htm HTTP/1.1`. We can see that the request type is `GET`, the document path is `/page1.htm`, and the protocol version is `HTTP/1.1`.

When dealing with text-based network protocols, it is always important to be explicit about line endings. This is because different operating systems have standardized on different line-ending conventions. Each line of an HTTP message ends with a carriage return, followed by a newline character. In C, this looks like `\r\n`. In practice, some web servers may tolerate other line endings. You should ensure that your clients always send a proper `\r\n` line ending for maximum compatibility.

After the request line, there are various HTTP header fields. Each header field consists of its name followed by a colon, and then its value. Consider the `User-Agent: Mozilla/5.0 (Windows NT 10.0; Win64; x64) AppleWebKit/537.36` line. This `User-Agent` line is telling the web server what software is contacting it. Some web servers will serve different documents to different user agents. For example, it is common for some websites to serve full documents to search engine spiders while serving paywalls to actual visitors. The server generally uses the user-agent HTTP header field to determine which is which. At the same time, there is a long history of web clients lying in the user-agent field. I suggest you take the high road in your applications and clearly identify your application with a unique user-agent value.

The only header field that is actually required is `Host`. The `Host` field tells the web server which web host the client is requesting the resource from. This is important because one web server may be hosting many different websites. The request line tells the web server that the `/page1.htm` document is wanted, but it doesn't specify which server that page is on. The `Host` field fills this role.

The `Connection: Keep-Alive` line tells the web server that the HTTP client would like to issue additional requests after the current request finishes. If the client had sent `Connection: Close` instead, that would indicate that the client intended to close the TCP connection after the HTTP response was received.

The web client must send a blank line after the HTTP request header. This blank line is how the web server knows that the HTTP request is finished. Without this blank line, the web server wouldn't know whether any additional header fields were still going to being sent. In C, the blank line looks like this: `\r\n\r\n`.

Let's now consider what the web server would send in reply to an HTTP request.

HTTP response format

Like the HTTP request, the HTTP response also consists of a header part and a body part. Also similar to the HTTP request, the body part is optional. Most HTTP responses do have a body part, though.

The server at `www.example.com` could respond to our HTTP request with the following reply:

```
HTTP/1.1 200 OK
Cache-Control: max-age=604800
Content-Type: text/html; charset=UTF-8
Date: Fri, 14 Dec 2018 16:46:09 GMT
```

```
Etag: "1541025663+gzip"
Expires: Fri, 21 Dec 2018 16:46:09 GMT
Last-Modified: Fri, 09 Aug 2013 23:54:35 GMT
Server: ECS (ord/5730)
Vary: Accept-Encoding
X-Cache: HIT
Content-Length: 1270

<!doctype html>
<html>
<head>
    <title>Example Domain</title>
...
```

The first line of an HTTP response is the **status line**. The status line consists of the protocol version, the response code, and the response code description. In the preceding example, we can see that the protocol version is HTTP/1.1, the response code is 200, and the response code description is OK. 200 OK is the typical response code to an HTTP GET request when everything goes ok. If the server couldn't find the resource the client has requested, it might respond with a 404 Page Not Found response code instead.

Many of the HTTP response headers are used to assist with caching. The Date, Etag, Expires, and Last-Modified fields can all be used by the client to cache documents.

The Content-Type field tells the client what type of resource it is sending. In the preceding example, it is an HTML web page, which is specified with text/html. HTTP can be used to send all types of resources, such as images, software, and videos. Each resource type has a specific Content-Type, which tells the client how to interpret the resource.

The Content-Length field specifies the size of the HTTP response body in bytes. In this case, we see that the requested resource is 1270 bytes long. There are a few ways to determine the body length, but the Content-Length field is the simplest. We will look at other ways in the *Response Body Length* section later in this chapter.

The HTTP response header section is delineated from the HTTP response body by a blank line. After this blank line, the HTTP body follows. Note that the HTTP body is not necessarily text-based. For example, if the client requested an image, then the HTTP body would likely be binary data. Also consider that, if the HTTP body is text-based, such as an HTML web page, it is free to use its own line-ending convention. It doesn't have to use the \r\n line ending required by HTTP.

If the client had sent a HEAD request type instead of `GET`, then the server would respond with exactly the same HTTP headers as before, but it would not include the HTTP body.

With the HTTP response format defined, let's look at some of the most common HTTP response types.

HTTP response codes

There are many different types of HTTP response codes.

If the request was successful, then the server responds with a code in the 200 range:

- `200 OK`: The client's request is successful, and the server sends the requested resource

If the resource has moved, the server can respond with a code in the 300 range. These codes are commonly used to redirect traffic from an unencrypted connection to an encrypted one, or to redirect traffic from a `www` subdomain to a bare one. They are also used if a website has undergone restructuring, but wants to keep incoming links working. The common 300 range codes are as follows:

- `301 Moved Permanently`: The requested resource has moved to a new location. This location is indicated by the server in the `Location` header field. All future requests for this resource should use this new location.
- `307 Moved Temporarily`: The requested resource has moved to a new location. This location is indicated by the server in the `Location` header field. This move may not be permanent, so future requests should still use the original location.

Errors are indicated by 400 or 500 range response codes. Some common ones are as follows:

- `400 Bad Request`: The server doesn't understand/support the client's request
- `401 Unauthorized`: The client isn't authorized for the requested resource
- `403 Forbidden`: The client is forbidden to access the requested resource
- `500 Internal Server Error`: The server encountered an error while trying to fulfill the client's request

In addition to a response type, the HTTP server must also be able to unambiguously communicate the length of the response body.

Response body length

The HTTP body response length can be determined a few different ways. The simplest is if the HTTP server includes a `Content-Length` header line in its response. In that case, the server simply states the body length directly.

If the server would like to begin sending data before the body's length is known, then it can't use the `Content-Length` header line. In this case, the server can send a `Transfer-Encoding: chunked` header line. This header line indicates to the client that the response body will be sent in separate chunks. Each chunk begins with its chunk length, encoded in base-16 (hexadecimal), followed by a newline, and then the chunk data. The entire HTTP body ends with a zero-length chunk.

Let's consider an HTTP response example that uses chunked encoding:

```
HTTP/1.1 200 OK
Content-Type: text/plain; charset=ascii
Transfer-Encoding: chunked

44
Lorem ipsum dolor sit amet, consectetur adipiscing elit, sed do eius
37
mod tempor incididunt ut labore et dolore magna aliqua.
0
```

In the preceding example, we see the HTTP body starts with `44` followed by a newline. This `44` should be interpreted as hexadecimal. We can use the built-in C `strtol()` function to interpret hexadecimal numbers.

Hexadecimal numbers are commonly written with a `0x` prefix to disambiguate them from decimal. We identify them with this prefix here, but keep in mind that the HTTP protocol does not add this prefix.

The `0x44` hexadecimal number is equal to 68 in decimal. After the `44` and newline, we see 68 characters that are part of the requested resource. After the 68 character chunk, the server sends a newline.

The server then sent `37`. `0x37` is 55 in decimal. After a newline, 55 characters are sent as chunk data. The server then sends a zero-length chunk to indicate that the response is finished.

The client should interpret the complete HTTP response after it has decoded the chunking as `Lorem ipsum dolor sit amet, consectetur adipiscing elit, sed do eiusmod tempor incididunt ut labore et dolore magna aliqua.`

There are a few other ways to indicate the HTTP response body length besides `Content-Length` and `Transfer-Encoding: chunked`. However, the server is limited to those two unless the client explicitly states support for additional encoding types in the HTTP request.

You may sometimes see a server that simply closes the TCP connection when it has finished transmitting a resource. That was a common way to indicate resource size in `HTTP/1.0`. However, this method shouldn't be used with `HTTP/1.1`. The issue with using a closed connection to indicate response length is that it's ambiguous as to why the connection was closed. It could be because all data has been sent, or it could be because of some other reason. Consider what would happen if a network cable is unplugged while data is being sent.

Now that we've seen the basics of HTTP requests and responses, let's look at how web resources are identified.

What's in a URL

Uniform Resource Locators (**URL**), also known as web addresses, provide a convenient way to specify particular web resources. You can navigate to a URL by typing it into your web browser's address bar. Alternately, if you're browsing a web page and click on a link, that link is indicated with a URL.

Consider the `http://www.example.com:80/res/page1.php?user=bob#account` URL. Visually, the URL can be broken down like this:

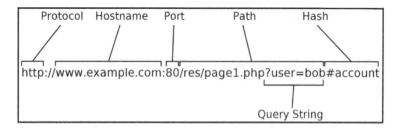

The URL can indicate the protocol, the host, the port number, the document path, and hash. However, the host is the only required part. The other parts can be implied.

We can parse the example URL from the preceding diagram:

- **http://**: The part before the first **://** indicates the protocol. In this example, the protocol is **http**, but it could be a different protocol such as `ftp://` or `https://`. If the protocol is omitted, the application will generally make an assumption. For example, your web browser would assume the protocol to be **http**.
- **www.example.com**: This specifies the hostname. It is used to resolve an IP address that the HTTP client can connect to. This hostname must also appear in the HTTP request `Host` header field. This is required since multiple hostnames can resolve to the same IP address. This part can also be an IP address instead of a name. IPv4 addresses are used directly (`http://192.168.50.1/`), but IPv6 addresses should be put inside square brackets (`http://[::1]/`).
- **:80**: The port number can be specified explicitly by using a colon after the hostname. If the port number is not specified, then the client uses the default port number for the given protocol. The default port number for **http** is **80**, and the default port number for **https** is **443**. Non-standard port numbers are common for testing and development.
- **/res/page1.php?user/bob**: This specifies the document path. The HTTP server usually makes a distinction between the part before and after the question mark, but the HTTP client should not assign significance to this. The part after the question mark is often called the query string.
- **#account**: This is called the hash. The hash specifies a position within a document, and the hash is not sent to the HTTP server. It instead allows a browser to scroll to a particular part of a document after the entire document is received from the HTTP server.

Now that we have a basic understanding of URLs, let's write code to parse them.

Parsing a URL

We will write a C function to parse a given URL.

The function takes as input a URL, and it returns as output the hostname, the port number, and the document path. To avoid needing to do manual memory management, the outputs are returned as pointers to specific parts of the input URL. The input URL is modified with terminating null pointers as required.

Our function begins by printing the input URL. This is useful for debugging. The code for that is as follows:

```
/*web_get.c excerpt*/

void parse_url(char *url, char **hostname, char **port, char** path) {
    printf("URL: %s\n", url);
```

The function then attempts to find :// in the URL. If found, it reads in the first part of the URL as a protocol. Our program only supports HTTP. If the given protocol is not HTTP, then an error is returned. The code for parsing the protocol is as follows:

```
/*web_get.c excerpt*/

    char *p;
    p = strstr(url, "://");

    char *protocol = 0;
    if (p) {
        protocol = url;
        *p = 0;
        p += 3;
    } else {
        p = url;
    }

    if (protocol) {
        if (strcmp(protocol, "http")) {
            fprintf(stderr,
                    "Unknown protocol '%s'. Only 'http' is supported.\n",
                    protocol);
            exit(1);
        }
    }
```

In the preceding code, a character pointer, p, is declared. protocol is also declared and set to 0 to indicate that no protocol has been found. strstr() is called to search for :// in the URL. If it is not found, then protocol is left at 0, and p is set to point back to the beginning of the URL. However, if :// is found, then protocol is set to the beginning of the URL, which contains the protocol. p is set to one after ://, which should be where the hostname begins.

If protocol was set, the code then checks that it points to the text http.

At this point in the code, `p` points to the beginning of the hostname. The code can save the hostname into the return variable, `hostname`. The code must then scan for the end of the hostname by looking for the first colon, slash, or hash. The code for this is as follows:

```
/*web_get.c excerpt*/

    *hostname = p;
    while (*p && *p != ':' && *p != '/' && *p != '#') ++p;
```

Once `p` has advanced to the end of the hostname, we must check whether a port number was found. A port number starts with a colon. If a port number is found, our code returns it in the `port` variable; otherwise, a default port number of `80` is returned. The code to check for a port number is as follows:

```
/*web_get.c excerpt*/

    *port = "80";
    if (*p == ':') {
        *p++ = 0;
        *port = p;
    }
    while (*p && *p != '/' && *p != '#') ++p;
```

After the port number, `p` points to the document path. The function returns this part of the URL in the `path` variable. Note that our function omits the first / in the path. This is for simplicity because it allows us to avoid allocating any memory. All document paths start with /, so the function caller can easily prepend that when the HTTP request is constructed.

The code to set the `path` variable is as follows:

```
/*web_get.c excerpt*/

    *path = p;
    if (*p == '/') {
        *path = p + 1;
    }
    *p = 0;
```

The code then attempts to find a hash, if it exists. If it does exist, it is overwritten with a terminating null character. This is because the hash is never sent to the web server and is ignored by our HTTP client.

The code that advances to the hash is as follows:

```
/*web_get.c excerpt*/

    while (*p && *p != '#') ++p;
    if (*p == '#') *p = 0;
```

Our function has now parsed out the hostname, port number, and document path. It then prints out these values for debugging purposes and returns. The final code for the parse_url() function is as follows:

```
/*web_get.c excerpt*/

    printf("hostname: %s\n", *hostname);
    printf("port: %s\n", *port);
    printf("path: %s\n", *path);
}
```

Now that we have code to parse a URL, we are one step closer to building an entire HTTP client.

Implementing a web client

We will now implement an HTTP web client. This client takes as input a URL. It then attempts to connect to the host and retrieve the resource given by the URL. The program displays the HTTP headers that are sent and received, and it attempts to parse out the requested resource content from the HTTP response.

Our program begins by including the chapter header, chap06.h:

```
/*web_get.c*/

#include "chap06.h"
```

We then define a constant, TIMEOUT. Later in our program, if an HTTP response is taking more than TIMEOUT seconds to complete, then our program abandons the request. You can define TIMEOUT as you like, but we give it a value of five seconds here:

```
/*web_get.c continued*/

#define TIMEOUT 5.0
```

Now, please include the entire `parse_url()` function as given in the previous section. Our client needs `parse_url()` to find the hostname, port number, and document path from a given URL.

Another helper function is used to format and send the HTTP request. We call it `send_request()`, and its code is given next:

```
/*web_get.c continued*/

void send_request(SOCKET s, char *hostname, char *port, char *path) {
    char buffer[2048];

    sprintf(buffer, "GET /%s HTTP/1.1\r\n", path);
    sprintf(buffer + strlen(buffer), "Host: %s:%s\r\n", hostname, port);
    sprintf(buffer + strlen(buffer), "Connection: close\r\n");
    sprintf(buffer + strlen(buffer), "User-Agent: honpwc web_get 1.0\r\n");
    sprintf(buffer + strlen(buffer), "\r\n");

    send(s, buffer, strlen(buffer), 0);
    printf("Sent Headers:\n%s", buffer);
}
```

`send_request()` works by first defining a character buffer in which to store the HTTP request. It then uses the `sprintf()` function to write to the buffer until the HTTP request is complete. The HTTP request ends with a blank line. This blank line tells the server that the entire request header has been received.

Once the request is formatted into `buffer`, `buffer` is sent over an open socket using `send()`. `buffer` is also printed to the console for debugging purposes. We define one more helper function for our web client. This function, `connect_to_host()`, takes in a hostname and port number and attempts to establish a new TCP socket connection to it.

In the first part of `connect_to_host()`, `getaddrinfo()` is used to resolve the hostname. `getnameinfo()` is then used to print out the server IP address for debugging purposes. The code for this is as follows:

```
/*web_get.c continued*/

SOCKET connect_to_host(char *hostname, char *port) {
    printf("Configuring remote address...\n");
    struct addrinfo hints;
    memset(&hints, 0, sizeof(hints));
    hints.ai_socktype = SOCK_STREAM;
    struct addrinfo *peer_address;
    if (getaddrinfo(hostname, port, &hints, &peer_address)) {
        fprintf(stderr, "getaddrinfo() failed. (%d)\n", GETSOCKETERRNO());
```

```
            exit(1);
    }

    printf("Remote address is: ");
    char address_buffer[100];
    char service_buffer[100];
    getnameinfo(peer_address->ai_addr, peer_address->ai_addrlen,
            address_buffer, sizeof(address_buffer),
            service_buffer, sizeof(service_buffer),
            NI_NUMERICHOST);
    printf("%s %s\n", address_buffer, service_buffer);
```

In the second part of `connect_to_host()`, a new socket is created with `socket()`, and a TCP connection is established with `connect()`. If everything goes well, the function returns the created socket. The code for the second half of `connect_to_host()` is as follows:

```
/*web_get.c continued*/

    printf("Creating socket...\n");
    SOCKET server;
    server = socket(peer_address->ai_family,
            peer_address->ai_socktype, peer_address->ai_protocol);
    if (!ISVALIDSOCKET(server)) {
        fprintf(stderr, "socket() failed. (%d)\n", GETSOCKETERRNO());
        exit(1);
    }

    printf("Connecting...\n");
    if (connect(server,
                peer_address->ai_addr, peer_address->ai_addrlen)) {
        fprintf(stderr, "connect() failed. (%d)\n", GETSOCKETERRNO());
        exit(1);
    }
    freeaddrinfo(peer_address);

    printf("Connected.\n\n");

    return server;
}
```

If you've been working through this book from the beginning, the code in `connect_to_host()` should be very familiar by now. If it's not, please refer to the previous chapters for a more detailed explanation of `getaddrinfo()`, `socket()`, and `connect()`. Chapter 3, *An In-Depth Overview of TCP Connections*, should be particularly helpful.

With our helper functions out of the way, we can now begin to define the `main()` function.
We begin `main()` with the following code:

```
/*web_get.c continued*/

int main(int argc, char *argv[]) {

#if defined(_WIN32)
    WSADATA d;
    if (WSAStartup(MAKEWORD(2, 2), &d)) {
        fprintf(stderr, "Failed to initialize.\n");
        return 1;
    }
#endif


    if (argc < 2) {
        fprintf(stderr, "usage: web_get url\n");
        return 1;
    }
    char *url = argv[1];
```

In the preceding code, Winsock is initialized, if needed, and the program's arguments are
checked. If a URL is given as an argument, it is stored in the `url` variable.

We can then parse the URL into its hostname, port, and path parts with the following code:

```
/*web_get.c continued*/

    char *hostname, *port, *path;
    parse_url(url, &hostname, &port, &path);
```

The program then continues by establishing a connection to the target server and sending
the HTTP request. This is made easy by using the two helper functions we defined
previously, `connect_to_host()` and `send_request()`. The code for this is as follows:

```
/*web_get.c continued*/

    SOCKET server = connect_to_host(hostname, port);
    send_request(server, hostname, port, path);
```

One feature of our web client is that it times out if a request takes too long to complete. In order to know how much time has elapsed, we need to record the start time. This is done using a call to the built-in `clock()` function. We store the start time in the `start_time` variable with the following:

```
/*web_get.c continued*/

    const clock_t start_time = clock();
```

It is now necessary to define some more variables that can be used for bookkeeping while receiving and parsing the HTTP response. The requisite variables are as follows:

```
/*web_get.c continued*/

#define RESPONSE_SIZE 8192
    char response[RESPONSE_SIZE+1];
    char *p = response, *q;
    char *end = response + RESPONSE_SIZE;
    char *body = 0;

    enum {length, chunked, connection};
    int encoding = 0;
    int remaining = 0;
```

In the preceding code, `RESPONSE_SIZE` is the maximum size of the HTTP response we reserve memory for. Our program is unable to parse HTTP responses bigger than this. If you extend this limit, it may be useful to use `malloc()` to reserve memory on the heap instead of the stack.

`response` is a character array that holds the entire HTTP response. `p` is a `char` pointer that keeps track of how far we have written into `response` so far. `q` is an additional `char` pointer that is used later. We define `end` as a `char` pointer, which points to the end of the `response` buffer. `end` is useful to ensure that we don't attempt to write past the end of our reserved memory.

The `body` pointer is used to remember the beginning of the HTTP response body once received.

If you recall, the HTTP response body length can be determined by a few different methods. We define an enumeration to list the method types, and we define the `encoding` variable to store the actual method used. Finally, the `remaining` variable is used to record how many bytes are still needed to finish the HTTP body or body chunk.

We then start a loop to receive and process the HTTP response. This loop first checks that it hasn't taken too much time and that we still have buffer space left to store the received data. The first part of this loop is as follows:

```
/*web_get.c continued*/

    while(1) {

        if ((clock() - start_time) / CLOCKS_PER_SEC > TIMEOUT) {
            fprintf(stderr, "timeout after %.2f seconds\n", TIMEOUT);
            return 1;
        }

        if (p == end) {
            fprintf(stderr, "out of buffer space\n");
            return 1;
        }
```

We then include the code to receive data over the TCP socket. Our code uses `select()` with a short timeout. This allows us to periodically check that the request hasn't timed out. You may recall from previous chapters that `select()` involves creating `fd_set` and `timeval` structures. The following code creates these objects and calls `select()`:

```
/*web_get.c continued*/

        fd_set reads;
        FD_ZERO(&reads);
        FD_SET(server, &reads);

        struct timeval timeout;
        timeout.tv_sec = 0;
        timeout.tv_usec = 200000;

        if (select(server+1, &reads, 0, 0, &timeout) < 0) {
            fprintf(stderr, "select() failed. (%d)\n", GETSOCKETERRNO());
            return 1;
        }
```

`select()` returns when either the timeout has elapsed, or new data is available to be read from the socket. Our code needs to use `FD_ISSET()` to determine whether new data is available to be read. If so, we read the data into the buffer at the `p` pointer.

Alternatively, when attempting to read new data, we may find that the socket was closed by the web server. If this is the case, we check whether we were expecting a closed connection to indicate the end of the transmission. That is the case if `encoding ==` `connection`. If so, we print the HTTP body data that was received.

The code for reading in new data and detecting a closed connection is as follows:

```
/*web_get.c continued*/

        if (FD_ISSET(server, &reads)) {
            int bytes_received = recv(server, p, end - p, 0);
            if (bytes_received < 1) {
                if (encoding == connection && body) {
                    printf("%.*s", (int)(end - body), body);
                }

                printf("\nConnection closed by peer.\n");
                break;
            }

            /*printf("Received (%d bytes): '%.*s'",
                    bytes_received, bytes_received, p);*/

            p += bytes_received;
            *p = 0;
```

Note that, in the preceding code, the `p` pointer is advanced to point to the end of received data. `*p` is set to zero, so our received data always ends with a null terminator. This allows us to use standard functions on the data that expect a null-terminated string. For example, we use the built-in `strstr()` function to search through the received data, and `strstr()` expects the input string to be null-terminated.

Next, if the HTTP body hasn't already been found, our code searches through the received data for a blank line that indicates the end of the HTTP header. A blank line is encoded by two consecutive line endings. HTTP defines a line ending as \r\n, so our code detects a blank line by searching for \r\n\r\n.

The following code finds the end of the HTTP header (which is the beginning of the HTTP body) using `strstr()`, and updates the `body` pointer to point to the beginning of the HTTP body:

```
/*web_get.c continued*/

            if (!body && (body = strstr(response, "\r\n\r\n"))) {
                *body = 0;
                body += 4;
```

It may be useful to print the HTTP header for debugging. This can be done with the following code:

```
/*web_get.c continued*/

            printf("Received Headers:\n%s\n", response);
```

Now that the headers have been received, we need to determine whether the HTTP server is using `Content-Length` or `Transfer-Encoding: chunked` to indicate body length. If it doesn't send either, then we assume that the entire HTTP body has been received once the connection is closed.

If the `Content-Length` is found using `strstr()`, we set `encoding = length` and store the body length in the `remaining` variable. The actual length is read from the HTTP header using the `strtol()` function.

If `Content-Length` is not found, then the code searches for `Transfer-Encoding: chunked`. If found, we set `encoding = chunked`. `remaining` is set to 0 to indicate that we haven't read in a chunk length yet.

If neither `Content-Length` or `Transfer-Encoding: chunked` is found, then `encoding = connection` is set to indicate that we consider the HTTP body received when the connection is closed.

The code for determining which body length method is used is as follows:

```
/*web_get.c continued*/

            q = strstr(response, "\nContent-Length: ");
            if (q) {
                encoding = length;
                q = strchr(q, ' ');
                q += 1;
                remaining = strtol(q, 0, 10);

            } else {
                q = strstr(response, "\nTransfer-Encoding: chunked");
                if (q) {
                    encoding = chunked;
                    remaining = 0;
                } else {
                    encoding = connection;
                }
            }
            printf("\nReceived Body:\n");
        }
```

The preceding code could be made more robust by doing case-insensitive searching, or by allowing for some flexibility in spacing. However, it should work with most web servers as is, and we're going to continue to keep it simple.

If the HTTP body start has been identified, and `encoding == length`, then the program simply needs to wait until `remaining` bytes have been received. The following code checks for this:

```
/*web_get.c continued*/

        if (body) {
            if (encoding == length) {
                if (p - body >= remaining) {
                    printf("%.*s", remaining, body);
                    break;
                }
```

In the preceding code, once `remaining` bytes of the HTTP body have been received, it prints the received body and breaks from the `while` loop.

If `Transfer-Encoding: chunked` is used, then the receiving logic is a bit more complicated. The following code handles this:

```
/*web_get.c continued*/

            } else if (encoding == chunked) {
                do {
                    if (remaining == 0) {
                        if ((q = strstr(body, "\r\n"))) {
                            remaining = strtol(body, 0, 16);
                            if (!remaining) goto finish;
                            body = q + 2;
                        } else {
                            break;
                        }
                    }
                    if (remaining && p - body >= remaining) {
                        printf("%.*s", remaining, body);
                        body += remaining + 2;
                        remaining = 0;
                    }
                } while (!remaining);
            }
        } //if (body)
```

In the preceding code, the `remaining` variable is used to indicate whether a chunk length or chunk data is expected next. When `remaining == 0`, the program is waiting to receive a new chunk length. Each chunk length ends with a newline; therefore, if a newline is found with `strstr()`, we know that the entire chunk length has been received. In this case, the chunk length is read using `strtol()`, which interprets the hexadecimal chunk length. `remaining` is set to the expected chunk length. A chunked message is terminated by a zero-length chunk, so if 0 was read, the code uses `goto finish` to break out of the main loop.

If the `remaining` variable is non-zero, then the program checks whether at least `remaining` bytes of data have been received. If so, that chunk is printed, and the `body` pointer is advanced to the end of the current chunk. This logic loops until it finds the terminating zero-length chunk or runs out of data.

At this point, we've shown all of the logic to parse the HTTP response body. We only need to end our loops, close the socket, and the program is finished. Here is the final code for `web_get.c`:

```
/*web_get.c continued*/

        } //if FDSET
    } //end while(1)
finish:

    printf("\nClosing socket...\n");
    CLOSESOCKET(server);

#if defined(_WIN32)
    WSACleanup();
#endif

    printf("Finished.\n");
    return 0;
}
```

You can compile and run `web_get.c` on Linux and macOS with the following commands:

```
gcc web_get.c -o web_get
./web_get http://example.com/
```

On Windows, the command to compile and run using MinGW is as follows:

```
gcc web_get.c -o web_get.exe -lws2_32
web_get.exe http://example.com/
```

Try running `web_get` with different URLs and study the outputs. You may find the HTTP response headers interesting.

The following screenshot shows what happens when we run `web_get` on `http://example.com/`:

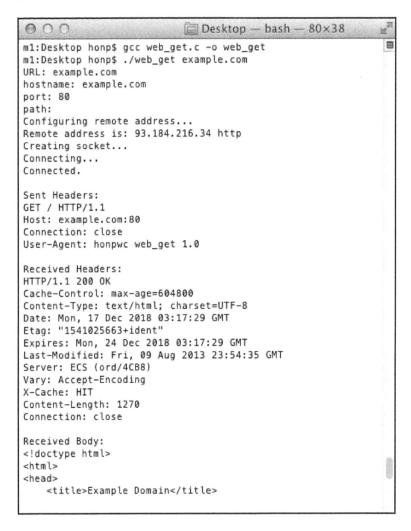

```
m1:Desktop honp$ gcc web_get.c -o web_get
m1:Desktop honp$ ./web_get example.com
URL: example.com
hostname: example.com
port: 80
path:
Configuring remote address...
Remote address is: 93.184.216.34 http
Creating socket...
Connecting...
Connected.

Sent Headers:
GET / HTTP/1.1
Host: example.com:80
Connection: close
User-Agent: honpwc web_get 1.0

Received Headers:
HTTP/1.1 200 OK
Cache-Control: max-age=604800
Content-Type: text/html; charset=UTF-8
Date: Mon, 17 Dec 2018 03:17:29 GMT
Etag: "1541025663+ident"
Expires: Mon, 24 Dec 2018 03:17:29 GMT
Last-Modified: Fri, 09 Aug 2013 23:54:35 GMT
Server: ECS (ord/4CB8)
Vary: Accept-Encoding
X-Cache: HIT
Content-Length: 1270
Connection: close

Received Body:
<!doctype html>
<html>
<head>
    <title>Example Domain</title>
```

`web_get` only supports the `GET` queries. `POST` queries are also common and useful. Let's now look at HTTP `POST` requests.

HTTP POST requests

An HTTP POST request sends data from the web client to the web server. Unlike an HTTP GET request, a POST request includes a body containing data (although this body could be zero-length).

The POST body format can vary, and it should be identified by a Content-Type header. Many modern, web-based APIs expect a POST body to be JSON encoded.

Consider the following HTTP POST request:

```
POST /orders HTTP/1.1
Host: example.com
User-Agent: Mozilla/5.0 (Windows NT 10.0; Win64; x64; rv:64.0)
Content-Type: application/json
Content-Length: 56
Connection: close

{"symbol":"VOO","qty":"10","side":"buy","type":"market"}
```

In the preceding example, you can see that the HTTP POST request is similar to an HTTP GET request. Notable differences are as follows: the request starts with POST instead of GET; a Content-Type header field is included; a Content-Length header field is present; and an HTTP message body is included. In that example, the HTTP message body is in JSON format, as specified by the Content-Type header.

Encoding form data

If you encounter a form on a website, such as a login form, that form likely uses a POST request to transmit its data to the web server. A standard HTML form encodes the data it sends in a format called **URL encoding**, also called **percent encoding**. When URL encoded form data is submitted in an HTTP POST request, it uses the Content-Type: application/x-www-form-urlencoded header.

Consider the following HTML for a submittable form:

```
<form method="post" action="/submission.php">

   <label for="name">Name:</label>
   <input name="name" type="text"><br>

   <label for="comment">Comment:</label>
   <input name="comment" type="text"><br>
```

```
    <input type="submit" value="submit">

</form>
```

In your web browser, the preceding HTML may render as shown in the following screenshot:

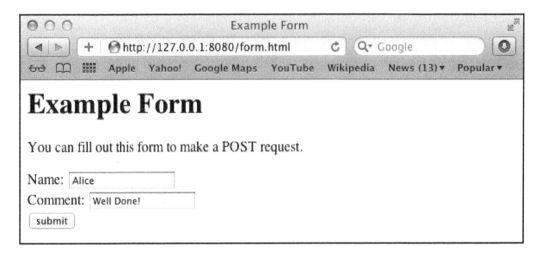

When this form is submitted, its data is encoded in an HTTP request such as the following:

```
POST /submission.php HTTP/1.1
Host: 127.0.0.1:8080
User-Agent: Mozilla/5.0 (Macintosh; Intel Mac OS X 10_7)
Accept-Language: en-US
Accept-Encoding: gzip, deflate
Content-Type: application/x-www-form-urlencoded
Content-Length: 31
Connection: keep-alive

name=Alice&comment=Well+Done%21
```

In the preceding HTTP request, you can see that `Content-Type: application/x-www-form-urlencoded` is used. In this format, each form field and value is paired by an equal sign, and multiple form fields are chained together by ampersands.

Special characters in form field names or values must be encoded. Notice that `Well Done!` was encoded as `Well+Done%21`. Spaces are encoded with plus symbols, and special characters are encoded by a percent sign followed by their two-digit hexadecimal value (thus, the exclamation point was encoded as `%21`). A percent sign itself would be encoded as `%25`.

File uploads

When an HTML form includes file uploads, the browser uses a different content type. In this case, `Content-Type: multipart/form-data` is used. When `Content-Type: multipart/form-data` is used, a boundary specifier is included. This boundary is a special delimiter, set by the sender, which separates parts of the submitted form data.

Consider the following HTML form:

```html
<form method="post" enctype="multipart/form-data" action="/submit.php">
    <input name="name" type="text"><br>
    <input name="comment" type="text"><br>
    <input name="file" type="file"><br>
    <input type="submit" value="submit">
</form>
```

If the user navigates to a web page bearing the HTML form from the preceding code, and enters the name `Alice`, the comment `Well Done!`, and selects a file to upload called `upload.txt`, then the following HTTP `POST` request could be sent by the browser:

```
POST /submit.php HTTP/1.1
Host: example.com
Content-Type: multipart/form-data; boundary=-----------233121195710604
Content-Length: 1727

-------------233121195710604
Content-Disposition: form-data; name="name"

Alice
-------------233121195710604
Content-Disposition: form-data; name="comment"

Well Done!
-------------233121195710604
Content-Disposition: form-data; name="file"; filename="upload.txt"
Content-Type: text/plain

Hello.... <truncated>
```

As you can see, when using `multipart/form-data`, each section of data is separated by a boundary. This boundary is what allows the receiver to delineate between separate fields or uploaded files. It is important that this boundary is chosen so that it does not appear in any submitted field or uploaded file!

Summary

HTTP is the protocol that powers the modern internet. It is behind every web page, every link click, every graphic loaded, and every form submitted. In this chapter, we saw that HTTP is a text-based protocol that runs over a TCP connection. We learned the HTTP formats for both client requests and server responses.

In this chapter, we also implemented a simple HTTP client in C. This client had a few non-trivial tasks – parsing a URL, formatting a GET request HTTP header, waiting for a response, and parsing the received data out of the HTTP response. In particular, we looked at handling two different methods of parsing out the HTTP body. The first, and easiest, method was Content-Length, where the entire body length is explicitly specified. The second method was chunked encoding, where the body is sent as separate chunks, which our program had to delineate between.

We also briefly looked at the POST requests and the content formats associated with them.

In the next chapter, Chapter 7, *Building a Simple Web Server*, we will develop the counterpart to our HTTP client—an HTTP server.

Questions

Try these questions to test your knowledge from this chapter:

1. Does HTTP use TCP or UDP?
2. What types of resources can be sent over HTTP?
3. What are the common HTTP request types?
4. What HTTP request type is typically used to send data from the server to the client?
5. What HTTP request type is typically used to send data from the client to the server?
6. What are the two common methods used to determine an HTTP response body length?
7. How is the HTTP request body formatted for a POST-type HTTP request?

The answers to these questions can be found in Appendix A, *Answers to Questions*.

Further reading

For more information about HTTP and HTML, please refer to the following resources:

- **RFC 7230**: *Hypertext Transfer Protocol (HTTP/1.1): Message Syntax and Routing* (`https://tools.ietf.org/html/rfc7230`)
- **RFC 7231**: *Hypertext Transfer Protocol (HTTP/1.1): Semantics and Content* (`https://tools.ietf.org/html/rfc7231`)
- **RFC 1866**: *Hypertext Markup Language – 2.0* (`https://tools.ietf.org/html/rfc1866`)
- **RFC 3986**: *Uniform Resource Identifier (URI): Generic Syntax* (`https://tools.ietf.org/html/rfc3986`)

Building a Simple Web Server 7

This chapter builds on the previous one by looking at the HTTP protocol from the server's perspective. In it, we will build a simple web server. This web server will work using the HTTP protocol, and you will be able to connect to it with any standard web browser. Although it won't be full-featured, it will be suitable for serving a few static files locally. It will be able to handle a few simultaneous connections from multiple clients at once.

The following topics are covered in this chapter:

- Accepting and buffering multiple connections
- Parsing an HTTP request line
- Formatting an HTTP response
- Serving a file
- Security considerations

Technical requirements

The example programs from this chapter can be compiled with any modern C compiler. We recommend MinGW on Windows and GCC on Linux and macOS; see Appendices B, *Setting Up Your C Compiler on Windows*, Appendices C, *Setting Up Your C Compiler on Linux*, and Appendices D, *Setting Up Your C Compiler on macOS*, for compiler setup.

The code for this book can be found at https://github.com/codeplea/Hands-On-Network-Programming-with-C.

From the command line, you can download the code for this chapter with the following command:

```
git clone https://github.com/codeplea/Hands-On-Network-Programming-with-C
cd Hands-On-Network-Programming-with-C/chap07
```

Each example program in this chapter runs on Windows, Linux, and macOS. While compiling on Windows, each example program requires linking to the **Winsock** library. This can be accomplished by passing the -lws2_32 option to gcc.

We provide the exact commands needed to compile each example as they are introduced.

All of the example programs in this chapter require the same header files and C macros that we developed in Chapter 2, *Getting to Grips with Socket APIs*. For brevity, we put these statements in a separate header file, chap07.h, which we can include in each program. For an explanation of these statements, please refer to Chapter 2, *Getting to Grips with Socket APIs*.

The content of chap07.h is as follows:

```c
/*chap07.h*/

#if defined(_WIN32)
#ifndef _WIN32_WINNT
#define _WIN32_WINNT 0x0600
#endif
#include <winsock2.h>
#include <ws2tcpip.h>
#pragma comment(lib, "ws2_32.lib")

#else
#include <sys/types.h>
#include <sys/socket.h>
#include <netinet/in.h>
#include <arpa/inet.h>
#include <netdb.h>
#include <unistd.h>
#include <errno.h>

#endif


#if defined(_WIN32)
#define ISVALIDSOCKET(s) ((s) != INVALID_SOCKET)
#define CLOSESOCKET(s) closesocket(s)
#define GETSOCKETERRNO() (WSAGetLastError())

#else
#define ISVALIDSOCKET(s) ((s) >= 0)
#define CLOSESOCKET(s) close(s)
#define SOCKET int
#define GETSOCKETERRNO() (errno)
#endif
```

```
#include <stdio.h>
#include <stdlib.h>
#include <string.h>
```

The HTTP server

In this chapter, we are going to implement an HTTP web server that can serve static files from a local directory. HTTP is a text-based client-server protocol that uses the **Transmission Control Protocol** (**TCP**).

When implementing our HTTP server, we need to support multiple, simultaneous connections from many clients at once. Each received **HTTP Request** needs to be parsed, and our server needs to reply with the proper **HTTP Response**. This **HTTP Response** should include the requested file if possible.

Consider the HTTP transaction illustrated in the following diagram:

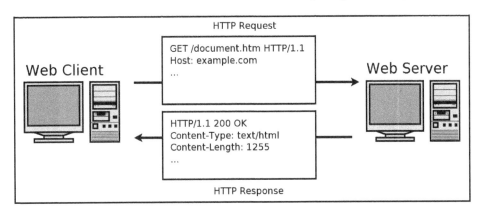

In the preceding diagram, the client is requesting /document.htm from the server. The server finds /document.htm and returns it to the client.

Our HTTP server is somewhat simplified, and we only need to look at the first line of the HTTP request. This first line is called the **request line**. Our server only supports GET type requests, so it needs to first check that the request line starts with GET. It then parses out the requested resource, /document.htm in the preceding example.

A more full-featured HTTP server would look at several other HTTP headers. It would look at the `Host` header to determine which site it is hosting. Our server only supports hosting one site, so this header is not meaningful for us.

A production server would also look at headers such as **Accept-Encoding** and **Accept-Language,** which could inform a proper response format. Our server just ignores these, and it instead serves files in only the most straightforward way.

The internet can sometimes be a hostile environment. A production-grade web server needs to include security in layers. It should be absolutely meticulous about file access and resource allocation. In the interest of clear explanation and brevity, the server we develop in this chapter is not security-hardened, and it should not be used on the public internet for this reason.

The server architecture

An HTTP server is a complicated program. It must handle multiple simultaneous connections, parse a complex text-based protocol, handle malformed requests with the proper errors, and serve files. The example we develop in this chapter is greatly simplified from a production-ready server, but it is still a few hundred lines of code. We benefit from breaking the program down into separate functions and data structures.

At the global level, our program stores a linked list of data structures. This linked list contains one separate data structure for each connected client. This data structure stores information about each client such as their address, their socket, and their data received so far. We implement many helper functions that work on this global linked list. These functions are used to add new clients, drop clients, wait on client data, look up clients by their socket (as sockets are returned by `select()`), serve files to clients, and send error messages to clients.

Our server's main loop can then be simplified. It waits for new connections or new data. When new data is received, it checks whether the data consists of a complete HTTP request. If a complete HTTP request is received, the server attempts to send the requested resource. If the HTTP request is malformed or the resource cannot be found, then the server sends an error message to the connected client instead.

Most of the server complexity lies in handling multiple connections, parsing the HTTP request, and handling error conditions.

The server is also responsible for telling the client the content type of each resource it sends. There are a few ways to accomplish this; let's consider them next.

Content types

It is the HTTP server's job to tell its client the type of content being sent. This is done by the `Content-Type` header. The value of the `Content-Type` header should be a valid media type (formerly known as the **MIME type**) registered with the **Internet Assigned Numbers Authority** (**IANA**). See the *Further reading* section of this chapter for a link to the IANA list of media types.

There are a few ways to determine the media type of a file. If you're on a Unix-based system, such as Linux or macOS, then your operating system already provides a utility for this.

Try the following command on Linux or macOS (replace `example.txt` with a real filename):

```
file --mime-type example.txt
```

The following screenshot shows its usage:

```
m1:chap07 honp$ file --mime-type public/index.html
public/index.html: text/html
m1:chap07 honp$ file --mime-type public/smile.png
public/smile.png: image/png
m1:chap07 honp$ file --mime-type public/test.txt
public/test.txt: text/plain
m1:chap07 honp$
```

As you can see in the preceding screenshot, the `file` utility told us the media type of `index.html` is `text/html`. It also said the media type of `smile.png` is `image/png`, and the media type of `test.txt` is `text/plain`.

Our web server just uses the file's extension to determine the media type.

Common file extensions and their media type are listed in the following table:

Extension	Media Type
.css	text/css
.csv	text/csv
.gif	image/gif
.htm	text/html
.html	text/html
.ico	image/x-icon
.jpeg	image/jpeg
.jpg	image/jpeg
.js	application/javascript
.json	application/json
.png	image/png
.pdf	application/pdf
.svg	image/svg+xml
.txt	text/plain

If a file's media type is unknown, then our server should use `application/octet-stream` as a default. This indicates that the browser should treat the content as an unknown binary blob.

Let's continue by writing code to get `Content-Type` from a filename.

Returning Content-Type from a filename

Our server code uses a series of `if` statements to determine the proper media type based only on the requested file's extension. This isn't a perfect solution, but it is a common one, and it works for our purposes.

The code to determine a file's media type is as follows:

```
/*web_server.c except*/

const char *get_content_type(const char* path) {
    const char *last_dot = strrchr(path, '.');
    if (last_dot) {
        if (strcmp(last_dot, ".css") == 0) return "text/css";
        if (strcmp(last_dot, ".csv") == 0) return "text/csv";
        if (strcmp(last_dot, ".gif") == 0) return "image/gif";
        if (strcmp(last_dot, ".htm") == 0) return "text/html";
        if (strcmp(last_dot, ".html") == 0) return "text/html";
```

```
        if (strcmp(last_dot, ".ico") == 0) return "image/x-icon";
        if (strcmp(last_dot, ".jpeg") == 0) return "image/jpeg";
        if (strcmp(last_dot, ".jpg") == 0) return "image/jpeg";
        if (strcmp(last_dot, ".js") == 0) return "application/javascript";
        if (strcmp(last_dot, ".json") == 0) return "application/json";
        if (strcmp(last_dot, ".png") == 0) return "image/png";
        if (strcmp(last_dot, ".pdf") == 0) return "application/pdf";
        if (strcmp(last_dot, ".svg") == 0) return "image/svg+xml";
        if (strcmp(last_dot, ".txt") == 0) return "text/plain";
    }

    return "application/octet-stream";
}
```

The `get_content_type()` function works by matching the filename extension to a list of known extensions. This is done by using the `strrchr()` function to find the last dot (.) in the filename. If a dot is found, then `strcmp()` is used to check for a match on each extension. When a match is found, the proper media type is returned. Otherwise, the default of `application/octet-stream` is returned instead.

Let's continue building helper functions for our server.

Creating the server socket

Before entertaining the exciting parts of the HTTP server, such as message parsing, let's get the basics out of the way. Our HTTP server, like all servers, needs to create a listening socket to accept new connections. We define a function, `create_socket()`, for this purpose. This function begins by using `getaddrinfo()` to find the listening address:

```
/*web_server.c except*/

SOCKET create_socket(const char* host, const char *port) {
    printf("Configuring local address...\n");
    struct addrinfo hints;
    memset(&hints, 0, sizeof(hints));
    hints.ai_family = AF_INET;
    hints.ai_socktype = SOCK_STREAM;
    hints.ai_flags = AI_PASSIVE;

    struct addrinfo *bind_address;
    getaddrinfo(host, port, &hints, &bind_address);
```

`create_socket()` then continues with creating a socket using `socket()`, binding that socket to the listening address with `bind()`, and having the socket enter a listening state with `listen()`. The following code calls these functions while detecting error conditions:

```
/*web_server.c except*/

    printf("Creating socket...\n");
    SOCKET socket_listen;
    socket_listen = socket(bind_address->ai_family,
            bind_address->ai_socktype, bind_address->ai_protocol);
    if (!ISVALIDSOCKET(socket_listen)) {
        fprintf(stderr, "socket() failed. (%d)\n", GETSOCKETERRNO());
        exit(1);
    }

    printf("Binding socket to local address...\n");
    if (bind(socket_listen,
                bind_address->ai_addr, bind_address->ai_addrlen)) {
        fprintf(stderr, "bind() failed. (%d)\n", GETSOCKETERRNO());
        exit(1);
    }
    freeaddrinfo(bind_address);

    printf("Listening...\n");
    if (listen(socket_listen, 10) < 0) {
        fprintf(stderr, "listen() failed. (%d)\n", GETSOCKETERRNO());
        exit(1);
    }

    return socket_listen;
}
```

The preceding code should be very familiar to you if you're working through this book in order. If not, please refer to `Chapter 3`, *An In-Depth Overview of TCP Connections*, for information about setting up TCP servers.

Multiple connections buffering

One important hurdle to overcome, when implementing any server software, is accepting and parsing requests from multiple clients simultaneously.

Consider a client that sends only the beginning of an HTTP request, followed by a delay, and then the remainder of the HTTP request. In this case, we cannot respond to that client until the entire HTTP request is received. However, at the same time, we do not wish to delay servicing other connected clients while waiting. For this reason, we need to buffer up received data for each client separately. Only once we've received an entire HTTP request from a client can we respond to that client.

It is useful to define a C `struct` to store information on each connected client. Our program uses the following:

```
/*web_server.c except*/

#define MAX_REQUEST_SIZE 2047

struct client_info {
    socklen_t address_length;
    struct sockaddr_storage address;
    SOCKET socket;
    char request[MAX_REQUEST_SIZE + 1];
    int received;
    struct client_info *next;
};
```

This `struct` allows us to store information about each connected client. A client's address is stored in the `address` field, the address length in `address_length`, and the socket in the `socket` field. All of the data received from the client so far is stored in the `request` array; `received` indicates the number of bytes stored in that array. The `next` field is a pointer that allows us to store `client_info` structures in a linked list.

To simplify our code, we store the root of our linked list in a global variable, `clients`. The declaration is as follows:

```
/*web_server.c except*/

static struct client_info *clients = 0;
```

Declaring `clients` as a global variable helps to keep our code slightly shorter and clearer. However, if you require the code to be re-entrant (for example, if you want multiple servers running simultaneously), you will want to avoid the global state. This can be done by passing around the linked list root as a separate argument to each function call. This chapter's code repository includes an example of this alternative technique in the `web_server2.c` file.

It is useful to define a number of helper functions that work on the `client_info` data structure and the `clients` linked list. We implement the following helper functions for these purposes:

- `get_client()` takes a SOCKET variable and searches our linked list for the corresponding `client_info` data structure.
- `drop_client()` closes the connection to a client and removes it from the `clients` linked list.
- `get_client_address()` returns a client's IP address as a string (character array).
- `wait_on_clients()` uses the `select()` function to wait until either a client has data available or a new client is attempting to connect.
- `send_400()` and `send_404()` are used to handle HTTP error conditions.
- `serve_resource()` attempts to transfer a file to a connected client.

Let's now implement these functions one at a time.

get_client()

Our `get_client()` function accepts a SOCKET and searches through the linked list of connected clients to return the relevant `client_info` for that SOCKET. If no matching `client_info` is found in the linked list, then a new `client_info` is allocated and added to the linked list. Therefore, `get_client()` serves two purposes—it can find an existing `client_info`, or it can create a new `client_info`.

`get_client()` takes a SOCKET as its input and return a `client_info` structure. The following code is the first part of the `get_client()` function:

```
/*web_server.c except*/

struct client_info *get_client(SOCKET s) {
    struct client_info *ci = clients;

    while(ci) {
        if (ci->socket == s)
            break;
        ci = ci->next;
    }

    if (ci) return ci;
```

In the preceding code, we created the `get_client()` function and implemented our linked list search functionality. First, the linked list root, `clients`, is saved into a temporary variable, `ci`. If `ci->socket` is the socket we are searching for, then the loop breaks and `ci` is returned. If the `client_info` structure for the given socket isn't found, then the code continues on and must create a new `client_info` structure. The following code achieves this:

```c
/*web_server.c except*/

    struct client_info *n =
        (struct client_info*) calloc(1, sizeof(struct client_info));

    if (!n) {
        fprintf(stderr, "Out of memory.\n");
        exit(1);
    }

    n->address_length = sizeof(n->address);
    n->next = clients;
    clients = n;
    return n;
}
```

In the preceding code, the `calloc()` function is used to allocate memory for a new `client_info` structure. The `calloc()` function also zeroes-out the data structure, which is useful in this case. The code then checks that the memory allocation succeeded, and it prints an error message if it fails.

The code then sets `n->address_length` to the proper size. This allows us to use `accept()` directly on the `client_info` address later, as `accept()` requires the maximum address length as an input.

The `n->next` field is set to the current global linked list root, and the global linked list root, `clients`, is set to `n`. This accomplishes the task of adding in the new data structure at the beginning of the linked list.

The `get_client()` function ends by returning the newly allocated `client_info` structure, `n`.

drop_client()

The `drop_client()` function searches through our linked list of clients and removes a given client.

The entire function is given in the following code:

```
/*web_server.c except*/

void drop_client(struct client_info *client) {
    CLOSESOCKET(client->socket);

    struct client_info **p = &clients;

    while(*p) {
        if (*p == client) {
            *p = client->next;
            free(client);
            return;
        }
        p = &(*p)->next;
    }

    fprintf(stderr, "drop_client not found.\n");
    exit(1);
}
```

As you can see in the preceding code, `CLOSESOCKET()` is first used to close and clean up the client's connection.

The function then declares a pointer-to-pointer variable, `p`, and sets it to `clients`. This pointer-to-pointer variable is useful because we can use it to change the value of `clients` directly. Indeed, if the client to be removed is the first element in the linked list, then `clients` needs to be updated, so that `clients` points to the second element in the list.

The code uses a `while` loop to walk through the linked list. Once it finds that `*p == client`, `*p`, is set to `client->next`, which effectively removes the client from the linked list, the allocated memory is then freed and the function returns.

Although `drop_client()` is a simple function, it is handy as it can be called in several circumstances. It must be called when we finish sending a client a resource, and it also must be called when we finish sending a client an error message.

get_client_address()

It is useful to have a helper function that converts a given client's IP address into text. This function is given in the following code snippet:

```
/*web_server.c except*/

const char *get_client_address(struct client_info *ci) {
    static char address_buffer[100];
    getnameinfo((struct sockaddr*)&ci->address,
            ci->address_length,
            address_buffer, sizeof(address_buffer), 0, 0,
            NI_NUMERICHOST);
    return address_buffer;
}
```

get_client_address() is a simple function. It first allocates a char array to store the IP address in. This char array is declared static, which ensures that its memory is available after the function returns. This means that we don't need to worry about having the caller free() the memory. The downside to this method is that get_client_address() has a global state and is not re-entrant-safe. See web_server2.c for an alternative version that is re-entrant-safe.

After a char buffer is available, the code simply uses getnameinfo() to convert the binary IP address into a text address; getnameinfo() was covered in detail in previous chapters, but Chapter 5, *Hostname Resolution and DNS*, has a particularly detailed explanation.

wait_on_clients()

Our server is capable of handling many simultaneous connections. This means that our server must have a way to wait for data from multiple clients at once. We define a function, wait_on_clients(), which blocks until an existing client sends data, or a new client attempts to connect. This function uses select() as described in previous chapters. Chapter 3, *An In-Depth Overview of TCP Connections*, has a detailed explanation of select().

The wait_on_clients() function is defined as follows:

```
/*web_server.c except*/

fd_set wait_on_clients(SOCKET server) {
    fd_set reads;
    FD_ZERO(&reads);
```

```
        FD_SET(server, &reads);
        SOCKET max_socket = server;

        struct client_info *ci = clients;

        while(ci) {
            FD_SET(ci->socket, &reads);
            if (ci->socket > max_socket)
                max_socket = ci->socket;
            ci = ci->next;
        }

        if (select(max_socket+1, &reads, 0, 0, 0) < 0) {
            fprintf(stderr, "select() failed. (%d)\n", GETSOCKETERRNO());
            exit(1);
        }

        return reads;
    }
```

In the preceding code, first a new `fd_set` is declared and zeroed-out. The server socket is then added to the `fd_set` first. Then the code loops through the linked list of connected clients and adds the socket for each one in turn. A variable, `max_socket`, is maintained throughout this process to store the maximum socket number as required by `select()`.

After all the sockets are added to `fd_set reads`, the code calls `select()`, and `select()` returns when one or more of the sockets in `reads` is ready.

The `wait_on_clients()` function returns `reads` so that the caller can see which socket is ready.

send_400()

In the case where a client sends an HTTP request which our server does not understand, it is helpful to send a code 400 error. Because errors of this type can arise in several situations, we wrap this functionality in the `send_400()` function. The entire function follows:

```
/*web_server.c except*/

void send_400(struct client_info *client) {
    const char *c400 = "HTTP/1.1 400 Bad Request\r\n"
        "Connection: close\r\n"
        "Content-Length: 11\r\n\r\nBad Request";
    send(client->socket, c400, strlen(c400), 0);
```

```
        drop_client(client);
}
```

The `send_400()` function first declares a text array with the entire HTTP response hard-coded. This text is sent using the `send()` function, and then the client is dropped by calling the `drop_client()` function we defined earlier.

send_404()

In addition to the `400 Bad Request` error, our server also needs to handle the case where a requested resource is not found. In this case, a `404 Not Found` error should be returned. We define a helper function to return this error as follows:

```
/*web_server.c except*/

void send_404(struct client_info *client) {
    const char *c404 = "HTTP/1.1 404 Not Found\r\n"
        "Connection: close\r\n"
        "Content-Length: 9\r\n\r\nNot Found";
    send(client->socket, c404, strlen(c404), 0);
    drop_client(client);
}
```

The `send_404()` function works exactly like the `send_400()` function defined previously.

serve_resource()

The `serve_resource()` function sends a connected client a requested resource. Our server expects all hosted files to be in a subdirectory called `public`. Ideally, our server should not allow access to any files outside of this `public` directory. However, as we shall see, enforcing this restriction may be more difficult than it first appears.

Our `serve_resource()` function takes as arguments a connected client and a requested resource path. The function begins as follows:

```
/*web_server.c except*/

void serve_resource(struct client_info *client, const char *path) {

    printf("serve_resource %s %s\n", get_client_address(client), path);
```

The connected client's IP address and the requested path are printed to aid in debugging. In a production server, you would also want to print additional information. Most production servers log the date, time, request method, the client's user-agent string, and the response code as a minimum.

Our function then normalizes the requested path. There are a few things to check for. First, if the path is /, then we need to serve a default file. There is a tradition of serving a file called index in that case, and, indeed, this is what our code does.

We also check that the path isn't too long. Once we ensure that the path is below a maximum length, we can use fixed-size arrays to store it without worrying about buffer overflows.

Our code also checks that the path doesn't contain two consecutive dots—. . . In file paths, two dots indicate a reference to a parent directory. However, for security reasons, we want to allow access only into our public directory. We do not want to provide access to any parent directory. If we allowed paths with . ., then a malicious client could send GET /../web_server.c HTTP/1.1 and gain access to our server source code!

The following code is used to redirect root requests and to prevent long or obviously malicious requests:

```
/*web_server.c except*/

    if (strcmp(path, "/") == 0) path = "/index.html";

    if (strlen(path) > 100) {
        send_400(client);
        return;
    }

    if (strstr(path, "..")) {
        send_404(client);
        return;
    }
```

Our code now needs to convert the path to refer to files in the public directory. This is done with the sprintf() function. First, a text-array is reserved, full_path, and then sprintf() is used to store the full path into it. We are able to reserve a fixed allocation for full_path, as our earlier code ensured that path does not exceed 100 characters in length.

The code to set `full_path` is as follows:

```
/*web_server.c except*/

    char full_path[128];
    sprintf(full_path, "public%s", path);
```

It is important to note that the directory separator differs between Windows and other operating systems. While Unix-based systems use a slash (/), Windows instead uses a backslash (\) as its standard. Many Windows functions handle the conversion automatically, but the difference is sometimes important. For our simple server, the slash conversion isn't an absolute requirement. However, we include it anyway as a good practice.

The following code converts slashes to backslashes on Windows:

```
/*web_server.c except*/

#if defined(_WIN32)
    char *p = full_path;
    while (*p) {
        if (*p == '/') *p = '\\';
        ++p;
    }
#endif
```

The preceding code works by stepping through the `full_path` text array and detecting slash characters. When a slash is found, it is simply overwritten with a backslash. Note that the C code `'\\'` is equivalent to only one backslash. This is because the backslash has special meaning in C, and therefore the first backslash is used to escape the second backslash.

At this point, our server can check whether the requested resource actually exists. This is done by using the `fopen()` function. If `fopen()` fails, for any reason, then our server assumes that the file does not exist. The following code sends a `404` error if the requested resource isn't available:

```
/*web_server.c except*/

    FILE *fp = fopen(full_path, "rb");

    if (!fp) {
        send_404(client);
        return;
    }
```

If `fopen()` succeeds, then we can use `fseek()` and `ftell()` to determine the requested file's size. This is important, as we need to use the file's size in the `Content-Length` header. The following code finds the file size and stores it in the `cl` variable:

```
/*web_server.c except*/

    fseek(fp, 0L, SEEK_END);
    size_t cl = ftell(fp);
    rewind(fp);
```

Once the file size is known, we also want to get the file's type. This is used in the `Content-Type` header. We already defined a function, `get_content_type()`, which makes this task easy. The content type is store in the variable `ct` by the following code:

```
/*web_server.c except*/

    const char *ct = get_content_type(full_path);
```

Once the file has been located and we have its length and type, the server can begin sending the HTTP response. We first reserve a temporary buffer to store header fields in:

```
/*web_server.c except*/

#define BSIZE 1024
    char buffer[BSIZE];
```

Once the buffer is reserved, the server prints relevant headers into it and then sends those headers to the client. This is done using `sprintf()` and then `send()` in turn. The following code sends the HTTP response header:

```
/*web_server.c except*/

    sprintf(buffer, "HTTP/1.1 200 OK\r\n");
    send(client->socket, buffer, strlen(buffer), 0);

    sprintf(buffer, "Connection: close\r\n");
    send(client->socket, buffer, strlen(buffer), 0);

    sprintf(buffer, "Content-Length: %u\r\n", cl);
    send(client->socket, buffer, strlen(buffer), 0);

    sprintf(buffer, "Content-Type: %s\r\n", ct);
    send(client->socket, buffer, strlen(buffer), 0);

    sprintf(buffer, "\r\n");
    send(client->socket, buffer, strlen(buffer), 0);
```

Note that the last `send()` statement sends `\r\n`. This has the effect of transmitting a blank line. This blank line is used by the client to delineate the HTTP header from the beginning of the HTTP body.

The server can now send the actual file content. This is done by calling `fread()` repeatedly until the entire file is sent:

```
/*web_server.c except*/

    int r = fread(buffer, 1, BSIZE, fp);
    while (r) {
        send(client->socket, buffer, r, 0);
        r = fread(buffer, 1, BSIZE, fp);
    }
```

In the preceding code, `fread()` is used to read enough data to fill `buffer`. This buffer is then transmitted to the client using `send()`. These steps are looped until `fread()` returns `0`; this indicates that the entire file has been read.

Note that `send()` may block on large files. In a truly robust, production-ready server, you would need to handle this case. It could be done by using `select()` to determine when each socket is ready to read. Another common method is to use `fork()` or similar APIs to create separate threads/processes for each connected client. For simplicity, our server accepts the limitation that `send()` blocks on large files. Please refer to Chapter 13, *Socket Programming Tips and Pitfalls*, for more information about the blocking behavior of `send()`.

The function can finish by closing the file handle and using `drop_client()` to disconnect the client:

```
/*web_server.c except*/

    fclose(fp);
    drop_client(client);
}
```

That concludes the `serve_resource()` function.

Keep in mind that while `serve_resource()` attempts to limit access to only the `public` directory, it is not adequate in doing so, and `serve_resource()` should not be used in production code without carefully considering additional access loopholes. We discuss more security concerns later in this chapter.

With these helper functions out of the way, implementing our main server loop is a much easier task. We begin that next.

The main loop

With our many helper functions out of the way, we can now finish `web_server.c`. Remember to first `#include chap07.h` and also add in all of the types and functions we've defined so far—`struct client_info`, `get_content_type()`, `create_socket()`, `get_client()`, `drop_client()`, `get_client_address()`, `wait_on_clients()`, `send_400()`, `send_404()`, and `serve_resource()`.

We can then begin the `main()` function. It starts by initializing Winsock on Windows:

```
/*web_server.c except*/

int main() {

#if defined(_WIN32)
    WSADATA d;
    if (WSAStartup(MAKEWORD(2, 2), &d)) {
        fprintf(stderr, "Failed to initialize.\n");
        return 1;
    }
#endif
```

We then use our earlier function, `create_socket()`, to create the listening socket. Our server listens on port `8080`, but feel free to change it. On Unix-based systems, listening on low port numbers is reserved for privileged accounts. For security reasons, our web server should be running with unprivileged accounts only. This is why we use `8080` as our port number instead of the HTTP standard port, `80`.

The code to create the server socket is as follows:

```
/*web_server.c except*/

    SOCKET server = create_socket(0, "8080");
```

If you want to accept connections from only the local system, and not outside systems, use the following code instead:

```
/*web_server.c except*/

    SOCKET server = create_socket("127.0.0.1", "8080");
```

We then begin an endless loop that waits on clients. We call `wait_on_clients()` to wait until a new client connects or an old client sends new data:

```
/*web_server.c except*/

    while(1) {

        fd_set reads;
        reads = wait_on_clients(server);
```

The `server` then detects whether a new `client` has connected. This case is indicated by `server` being set in `fd_set reads`. We use the `FD_ISSET()` macro to detect this condition:

```
/*web_server.c except*/

        if (FD_ISSET(server, &reads)) {
            struct client_info *client = get_client(-1);

            client->socket = accept(server,
                    (struct sockaddr*) &(client->address),
                    &(client->address_length));

            if (!ISVALIDSOCKET(client->socket)) {
                fprintf(stderr, "accept() failed. (%d)\n",
                        GETSOCKETERRNO());
                return 1;
            }


            printf("New connection from %s.\n",
                    get_client_address(client));
        }
```

Once a new client connection has been detected, `get_client()` is called with the argument −1; −1 is not a valid socket specifier, so `get_client()` creates a new `struct client_info`. This `struct client_info` is assigned to the `client` variable.

The `accept()` socket function is used to accept the new connection and place the connected clients address information into the respective `client` fields. The new socket returned by `accept()` is stored in `client->socket`.

The client's address is printed using a call to `get_client_address()`. This is helpful for debugging.

Our server must then handle the case where an already connected client is sending data. This is a bit more complicated. We first walk through the linked list of clients and use `FD_ISSET()` on each client to determine which clients have data available. Recall that the linked list root is stored in the `clients` global variable.

We begin our linked list walk with the following:

```
/*web_server.c except*/

        struct client_info *client = clients;
        while(client) {
            struct client_info *next = client->next;

            if (FD_ISSET(client->socket, &reads)) {
```

We then check that we have memory available to store more received data for `client`. If the client's buffer is already completely full, then we send a `400` error. The following code checks for this condition:

```
/*web_server.c except*/

            if (MAX_REQUEST_SIZE == client->received) {
                send_400(client);
                continue;
            }
```

Knowing that we have at least some memory left to store received data, we can use `recv()` to store the client's data. The following code uses `recv()` to write new data into the client's buffer while being careful to not overflow that buffer:

```
/*web_server.c except*/

            int r = recv(client->socket,
                client->request + client->received,
                MAX_REQUEST_SIZE - client->received, 0);
```

A client that disconnects unexpectedly causes `recv()` to return a non-positive number. In this case, we need to use `drop_client()` to clean up our memory allocated for that client:

```
/*web_server.c except*/

            if (r < 1) {
                printf("Unexpected disconnect from %s.\n",
                    get_client_address(client));
                drop_client(client);
```

If the received data was written successfully, our server adds a null terminating character to the end of that client's data buffer. This allows us to use `strstr()` to search the buffer, as the null terminator tells `strstr()` when to stop.

Recall that the HTTP header and body is delineated by a blank line. Therefore, if `strstr()` finds a blank line (\r\n\r\n), we know that the HTTP header has been received and we can begin to parse it. The following code detects whether the HTTP header has been received:

```
/*web_server.c except*/

            } else {
                client->received += r;
                client->request[client->received] = 0;

                char *q = strstr(client->request, "\r\n\r\n");
                if (q) {
```

Our server only handles GET requests. We also enforce that any valid path should start with a slash character; `strncmp()` is used to detect these two conditions in the following code:

```
/*web_server.c except*/

                if (strncmp("GET /", client->request, 5)) {
                    send_400(client);
                } else {
                    char *path = client->request + 4;
                    char *end_path = strstr(path, " ");
                    if (!end_path) {
                        send_400(client);
                    } else {
                        *end_path = 0;
                        serve_resource(client, path);
                    }
                }
            } //if (q)
```

In the preceding code, a proper GET request causes the execution of the `else` branch. Here, we set the `path` variable to the beginning of the request path, which is starting at the fifth character of the HTTP request (because C arrays start at zero, the fifth character is located at `client->request + 4`).

The end of the requested path is indicated by finding the next space character. If found, we just call our `serve_resource()` function to fulfil the client's request.

Our server is basically functional at this point. We only need to finish our loops and close out the `main()` function. The following code accomplishes this:

```
/*web_server.c except*/

                    }
                }

            client = next;
        }

    } //while(1)


    printf("\nClosing socket...\n");
    CLOSESOCKET(server);


#if defined(_WIN32)
    WSACleanup();
#endif

    printf("Finished.\n");
    return 0;
}
```

Note that our server doesn't actually have a way to break from its infinite loop. It simply listens to connections forever. As an exercise, you may want to add in functionality that allows the server to shut down cleanly. This was omitted only to keep the code simpler. It may also be useful to drop all connected clients with this line of code—while(clients) drop_client(clients);

That concludes the code for web_server.c. I recommend you download web_server.c from this book's code repository and try it out.

You can compile and run web_server.c on Linux and macOS with the following commands:

```
gcc web_server.c -o web_server
./web_server
```

On Windows, the command to compile and run using MinGW is as follows:

```
gcc web_server.c -o web_server.exe -lws2_32
web_server.exe
```

The following screenshot shows the server being compiled and run on macOS:

```
● ○ ○                chap07 — web_server — 80×15
m1:chap07 honp$ gcc web_server.c -o web_server
m1:chap07 honp$ ./web_server
Configuring local address...
Creating socket...
Binding socket to local address...
Listening...
New connection from 127.0.0.1.
serve_resource 127.0.0.1 /index.html
New connection from 127.0.0.1.
serve_resource 127.0.0.1 /smile.png
```

If you connect to the server using a standard web browser, you should see something such as the following screenshot:

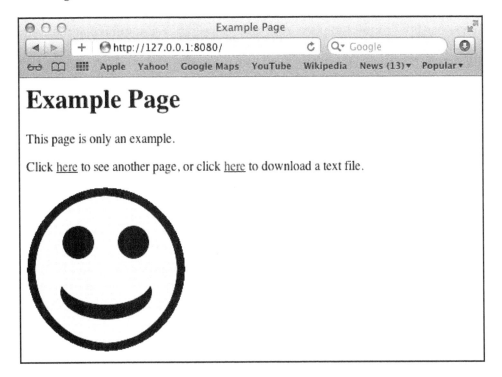

You can also drop different files into the `public` folder and play around with creating more complicated websites.

An alternative source file, `web_server2.c`, is also provided in this chapter's code repository. It behaves exactly like the code we developed, but it avoids having global state (at the expense of a little added verbosity). This may make `web_server2.c` more suitable for integration into more significant projects and continued development.

Although the web server we developed certainly works, it does have a number of shortcomings. Please don't deploy this server (or any other network code) in the wild without very carefully considering these shortcomings, some of which we address next.

Security and robustness

One of the most important rules, when developing networked code, is that your program should never trust the connected peer. Your code should never assume that the connected peer sends data in a particular format. This is especially vital for server code that may communicate with multiple clients at once.

If your code doesn't carefully check for errors and unexpected conditions, then it will be vulnerable to exploits.

Consider the following code which receives data into a buffer until a **space** character is found:

```
char buffer[1028] = {0};
char *p = buffer;

while (!strstr(p, " "))
    p += recv(client, p, 1028, 0);
```

The preceding code works simply. It reserves 1,028 bytes of buffer space and then uses `recv()` to write received data into that space. The p pointer is updated on each read to indicate where the next data should be written. The code then loops until the `strstr()` function detects a space character.

That code could be useful to read data from a client until an HTTP verb is detected. For example, it could receive data until GET is received, at which point the server can begin to process a GET request.

One problem with the preceding code is that `recv()` could write past the end of the allocated space for `buffer`. This is because `1028` is passed to `recv()`, even if some data has already been written. If a network client can cause your code to write past the end of a buffer, then that client may be able to completely compromise your server. This is because both data and executable code are stored in your server's memory. A malicious client may be able to write executable code past the `buffer` array and cause your program to execute it. Even if the malicious code isn't executed, the client could still overwrite other important data in your server's memory.

The preceding code can be fixed by passing to `recv()` only the amount of buffer space remaining:

```
char buffer[1028] = {0};
char *p = buffer;

while (!strstr(p, " "))
    p += recv(client, p, 1028 - (p - buffer), 0);
```

In this case, `recv()` is not be able to write more than 1,028 bytes total into `buffer`. You may think that the memory errors are resolved, but you would still be wrong. Consider a client that sends 1,028 bytes, but no space characters. Your code then calls `strstr()` looking for a space character. Considering that `buffer` is completely full now, `strstr()` cannot find a space character or a null terminating character! In that case, `strstr()` continues to read past the end of `buffer` into unallocated memory.

So, you fix this issue by only allowing `recv()` to write 1,027 bytes total. This reserves one byte to remain as the null terminating character:

```
char buffer[1028] = {0};
char *p = buffer;

while (!strstr(p, " "))
    p += recv(client, p, 1027 - (p - buffer), 0);
```

Now your code won't write or read past the array bounds for `buffer`, but the code is still very broken. Consider a client that sends 1,027 characters. Or consider a client that sends a single null character. In either case, the preceding code continues to loop forever, thus locking up your server and preventing other clients from being served.

Hopefully, the previous examples illustrate the care needed to implement a server in C. Indeed, it's easy to create bugs in any programming language, but in C special care needs to be taken to avoid memory errors.

Another issue with server software is that the server wants to allow access to some files on the system, but not others. A malicious client could send an HTTP request that tries to download arbitrary files from your server system. For example, if an HTTP request such as `GET /../secret_file.c HTTP/1.1` was sent to a naive HTTP server, that server may send the `secret_file.c` to the connected client, even though it exists outside of the `public` directory!

Our code in `web_server.c` detects the most obvious attempts at this by searching for requests containing `..` and denying those requests.

A robust server should use operating systems features to detect that requested files exist as actual files in the permitted directory. Unfortunately, there is no cross-platform way to do this, and the platform-dependent options are somewhat complicated.

Please understand that these are not purely theoretical concerns, but actual exploitable bugs. For example, if you run our `web_server.c` program on Windows and a client sends the request `GET /this_will_be_funny/PRN HTTP/1.1`, what do you suppose happens?

The `this_will_be_funny` directory doesn't exist, and the `PRN` file certainly doesn't exist in that non-existent directory. These facts may lead you to think that the server simply returns a `404 Not Found` error, as expected. However, that's not what happens. Under Windows, `PRN` is a special filename. When your server calls `fopen()` on this special name, Windows doesn't look for a file, but rather it connects to a *printer* interface! Other special names include `COM1` (connects to serial port 1) and `LPT1` (connects to parallel port 1), although there are others. Even if these filenames have an extension, such as `PRN.txt`, Windows still redirects instead of looking for a file.

One generally applicable piece of security advice is this—run your networked programs under non-privileged accounts that have access to only the minimum resources needed to function. In other words, if you are going to run a networked server, create a new account to run it under. Give that account read access to only the files that server needs to serve. This is not a substitute for writing secure code, but rather running as a non-privilege user creates one final barrier. It is advice you should apply even when running hardened, industry-tested server software.

Hopefully, the previous examples illustrate that programming is complicated, and safe network programming in C can be difficult. It is best approached with care. Oftentimes, it is not possible to know that you have all the loopholes covered. Operating systems don't always have adequate documentation. Operating system APIs often behave in non-obvious and non-intuitive ways. Be careful.

Open source servers

The code developed in this chapter is suitable for use in trusted applications on trusted networks. For example, if you are developing a video game, it can be very useful to make it serve a web page that displays debugging information. This doesn't have to be a security concern, as it can limit connections to the local machine.

If you must deploy a web server on the internet, I suggest you consider using a free and open source implementation that's already available. The web servers Nginx and Apache, for example, are highly performant, cross-platform, secure, written in C, and completely free. They are also well-documented and easy to find support for.

If you want to expose your program to the internet, you can communicate to a web server using either CGI or FastCGI. With CGI, the web server handles the HTTP request. When a request comes in, it runs your program and returns your program's output in the HTTP response body.

Alternatively, many web servers (such as Nginx or Apache) work as a reverse proxy. This essentially puts the web server between your code and the internet. The web server accepts and forwards HTTP messages to your HTTP server. This can have the effect of slightly shielding your code from attackers.

Summary

In this chapter, we worked through implementing an HTTP server in C from scratch. That's no small feat! Although the text-based nature of HTTP makes parsing HTTP requests simple, we needed to spend a lot of effort to ensure that multiple clients could be served simultaneously. We accomplished this by buffering received data for each client separately. Each client's state information was organized into a linked list.

Another difficulty was ensuring the safe handling of received data and detecting errors. We learned that a programmer must be very careful when handling network data to avoid creating security risks. We also saw that even very subtle issues, such as Windows's special filenames, can potentially create dangerous security holes for networked server applications.

In the next chapter, Chapter 8, *Making Your Program Send Email*, we move on from HTTP and consider the primary protocol associated with email—**Simple Mail Transfer Protocol (SMTP)**.

Questions

Try these questions to test your knowledge from this chapter:

1. How does an HTTP client indicate that it has finished sending the HTTP request?
2. How does an HTTP client know what type of content the HTTP server is sending?
3. How can an HTTP server identify a file's media type?
4. How can you tell whether a file exists on the filesystem and is readable by your program? Is `fopen(filename, "r") != 0` a good test?

The answers to these questions can be found in `Appendix A`, *Answers to Questions*.

Further reading

For more information about HTTP and HTML, please refer to the following:

- **RFC 7230**: *Hypertext Transfer Protocol (HTTP/1.1): Message Syntax and Routing* (`https://tools.ietf.org/html/rfc7230`)
- **RFC 7231**: *Hypertext Transfer Protocol (HTTP/1.1): Semantics and Content* (`https://tools.ietf.org/html/rfc7231`)
- *Media Types* (`https://www.iana.org/assignments/media-types/media-types.xhtml`)

8
Making Your Program Send Email

In this chapter, we will consider the protocol responsible for delivering email on the internet. This protocol is called the **Simple Mail Transfer Protocol** (**SMTP**).

Following an exposition of the inner workings of email transfer, we will build a simple SMTP client capable of sending short emails.

The following topics are covered in this chapter:

- How SMTP servers work
- Determining which mail server is responsible for a given domain
- Using SMTP
- Email encoding
- Spam-blocking and email-sending pitfalls
- SPF, DKIM, and DMARC

Technical requirements

The example programs from this chapter can be compiled with any modern C compiler. We recommend MinGW on Windows and GCC on Linux and macOS. See `Appendix B`, *Setting Up Your C Compiler on Windows*, `Appendix C`, *Setting Up Your C Compiler on Linux*, and `Appendix D`, *Setting Up Your C Compiler on macOS*, for compiler setup.

The code for this book can be found at `https://github.com/codeplea/Hands-On-Network-Programming-with-C`.

From the command line, you can download the code for this chapter with the following command:

```
git clone https://github.com/codeplea/Hands-On-Network-Programming-with-C
cd Hands-On-Network-Programming-with-C/chap08
```

Each example program in this chapter runs on Windows, Linux, and macOS. When compiling on Windows, each example program requires linking with the Winsock library. This can be accomplished by passing the `-lws2_32` option to `gcc`.

We provide the exact commands needed to compile each example as they are introduced.

All of the example programs in this chapter require the same header files and C macros that we developed in `Chapter 2`, *Getting to Grips with Socket APIs*. For brevity, we put these statements in a separate header file, `chap08.h`, which we can include in each program. For a detailed explanation of these statements, please refer to `Chapter 2`, *Getting to Grips with Socket APIs*.

The first part of `chap08.h` includes the needed networking headers for each platform. The code for this is as follows:

```
/*chap08.h*/

#if defined(_WIN32)
#ifndef _WIN32_WINNT
#define _WIN32_WINNT 0x0600
#endif
#include <winsock2.h>
#include <ws2tcpip.h>
#pragma comment(lib, "ws2_32.lib")

#else
#include <sys/types.h>
#include <sys/socket.h>
#include <netinet/in.h>
#include <arpa/inet.h>
#include <netdb.h>
#include <unistd.h>
#include <errno.h>

#endif
```

We also define some macros to make writing portable code easier, and we'll include the additional headers that our programs require:

```
/*chap08.h continued*/

#if defined(_WIN32)
#define ISVALIDSOCKET(s) ((s) != INVALID_SOCKET)
#define CLOSESOCKET(s) closesocket(s)
#define GETSOCKETERRNO() (WSAGetLastError())

#else
#define ISVALIDSOCKET(s) ((s) >= 0)
#define CLOSESOCKET(s) close(s)
#define SOCKET int
#define GETSOCKETERRNO() (errno)
#endif


#include <stdio.h>
#include <stdlib.h>
#include <string.h>
```

That concludes `chap08.h`.

Email servers

SMTP is the protocol responsible for delivering emails between servers. It is a text-based protocol operating on TCP port 25.

Not all emails need to be delivered between systems. For example, imagine you have a Gmail account. If you compose and send an email to your friend who also has a Gmail account, then SMTP is not necessarily used. In this case, Gmail only needs to copy your email into their inbox (or do equivalent database updates).

On the other hand, consider a case where you send an email to your friend's Yahoo! account. If the email is sent from your Gmail account, then it's clear that the Gmail and Yahoo! servers must communicate. In that case, your email is transmitted from the Gmail server to the Yahoo! server using SMTP.

This connection is illustrated in the following diagram:

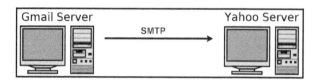

Retrieving your email from your mail service provider is a different issue than delivering email between service providers. Webmail is very popular now for sending and receiving mail from your mail provider. Webmail providers allow mailbox access through a web browser. Web browsers communicate using HTTP (or HTTPS).

Let's consider the full path of an email from **Alice** to **Bob**. In this example, Alice has **Gmail** as her mail provider, and Bob has **Yahoo!** as his mail provider. Both Alice and Bob access their mailbox using a standard web browser. The path an email takes from Bob to Alice is illustrated by the following diagram:

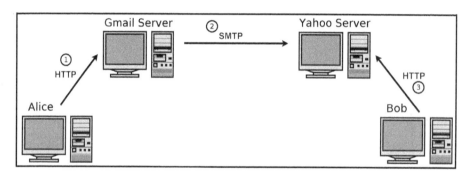

As you can see in the preceding diagram, **SMTP** is only used when delivering the mail between mail providers. This usage of SMTP is called **mail transmission**.

In fact, the email in this example could take other paths. Let's consider that Alice uses a desktop email client instead of webmail. Gmail still supports desktop email clients, and these clients offer many good features, even if they are falling out of fashion. A typical desktop client connects to a mail provider using either: **Internet Message Access Protocol (IMAP)** or **Post Office Protocol (POP)** and SMTP. In this case, SMTP is used by Alice to deliver her mail to her mail provider (Gmail). This usage of SMTP is called **mail submission**.

The Gmail provider then uses SMTP again to deliver the email to the **Yahoo!** mail server. This is illustrated in the following diagram:

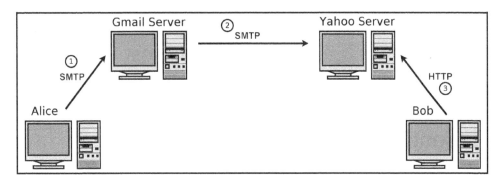

In the preceding diagram, the Gmail server would be considered an **SMTP relay**. In general, an SMTP server should only relay mail for trusted users. If an SMTP server relayed all mail, it would become quickly overwhelmed by spammers.

Many mail providers have a set of mail servers used to accept incoming mail and a separate set of mail servers used to accept outgoing mail from their users.

It is important to understand that SMTP is used to send mail. SMTP is not used to retrieve mail from a server. IMAP and POP are the common protocols used by desktop mail programs to retrieve mail from a server.

It is not necessary for Alice to send her mail through her provider's SMTP server. Instead, she could send it directly to Bob's mail provider as illustrated in the following diagram:

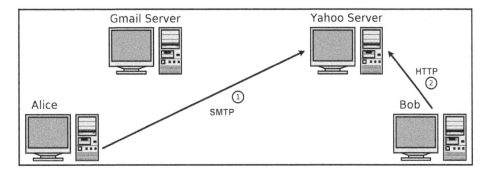

In practice, one usually delegates the delivery responsibility to their mail provider. This has several advantages; namely, the mail provider can attempt a redelivery if the destination mail server isn't available. Other advantages are discussed later in this chapter.

The program we develop in this chapter is used to deliver mail directly to the recipient's email provider. It is not useful to deliver mail to a relay server because we are not going to implement authentication techniques. Generally, SMTP servers do not relay mail without authenticating that the sender has an account with them.

The SMTP protocol we describe in this chapter is unsecured and not encrypted. This is convenient for explanation and learning purposes, but in the real world, you may want to secure your email transfer.

SMTP security

We describe unsecured SMTP in this chapter. In real-world use, SMTP should be secured if both communicating servers support it. Not all do.

Securing SMTP is done by having SMTP connections start out as plaintext on port 25. The SMTP client then issues a STARTTLS command to upgrade to a secure, encrypted connection. This secure connection works by merely running the SMTP commands through a TLS layer; therefore, everything we cover in this chapter is applicable to secure SMTP too. See Chapter 9, *Loading Secure Web Pages with HTTPS and OpenSSL,* for more information about TLS.

Mail transmission between servers is always done on port 25.

Many desktop email clients use TCP ports 465 or 587 for SMTP mail submission. **Internet Service Providers (ISPs)** prefer these alternative ports for mail submission, and it allows them to block port 25 altogether. This is generally justified as a spam prevention technique.

Next, let's see how to determine which mail server receives mail for a given email address.

Finding an email server

Consider the email address bob@example.com. In this case, bob identifies the user, while example.com identifies the domain name of the service provider. These parts are delineated by the @ symbol.

One domain name can potentially use multiple mail servers. Likewise, one mail server can provide service for many domain names. For this reason, identifying the mail server or servers responsible for receiving mail for `bob@example.com` isn't as easy as connecting to `example.com`. Instead, the mail server must be identified by performing a DNS lookup for an MX record.

DNS was covered in depth back in `Chapter 5`, *Hostname Resolution and DNS*. The program we developed in that chapter can be used to query MX records.

Otherwise, most operating systems provide a command-line tool for DNS lookup. Windows provides `nslookup`, while Linux and macOS provide `dig`.

On Windows, we can find the mail servers responsible for accepting mail `@gmail.com` using the following command:

```
nslookup -type=mx gmail.com
```

This lookup is shown in the following screenshot:

On Linux or macOS, an MX record lookup for a `@gmail.com` account is done with the following command:

```
dig mx gmail.com
```

The use of `dig` is shown in the following screenshot:

```
● ○ ○                     📁 Desktop — bash — 84×24
m1:Desktop honp$ dig mx gmail.com

; <<>> DiG 9.7.3 <<>> mx gmail.com
;; global options: +cmd
;; Got answer:
;; ->>HEADER<<- opcode: QUERY, status: NOERROR, id: 18189
;; flags: qr rd ra; QUERY: 1, ANSWER: 5, AUTHORITY: 0, ADDITIONAL: 0

;; QUESTION SECTION:
;gmail.com.                    IN      MX

;; ANSWER SECTION:
gmail.com.           5         IN      MX      5 gmail-smtp-in.l.google.com.
gmail.com.           5         IN      MX      10 alt1.gmail-smtp-in.l.google.com.
gmail.com.           5         IN      MX      20 alt2.gmail-smtp-in.l.google.com.
gmail.com.           5         IN      MX      30 alt3.gmail-smtp-in.l.google.com.
gmail.com.           5         IN      MX      40 alt4.gmail-smtp-in.l.google.com.

;; Query time: 51 msec
;; SERVER: 192.168.182.2#53(192.168.182.2)
;; WHEN: Tue Jan 15 15:17:09 2019
;; MSG SIZE  rcvd: 150

m1:Desktop honp$ ▌
```

As you can see in the preceding two screenshots, Gmail uses five mail servers. When multiple MX records are found, mail should be delivered to the server having the lowest MX preference first. If mail delivery fails to that server, then an attempt should be made to deliver to the server having the next lowest preference, and so on. At the time of this writing, Gmail's primary mail server, having a preference of 5, is `gmail-smtp-in.l.google.com`. That is the SMTP server you would connect to in order to send mail to an `@gmail.com` address.

It is also possible for MX records to have the same preference. Yahoo! uses mail servers having the same preference. The following screenshot shows the MX records for `yahoo.com`:

```
● ○ ○                    📁 Desktop — bash — 84×24
m1:Desktop honp$ dig mx yahoo.com

; <<>> DiG 9.7.3 <<>> mx yahoo.com
;; global options: +cmd
;; Got answer:
;; ->>HEADER<<- opcode: QUERY, status: NOERROR, id: 30161
;; flags: qr rd ra; QUERY: 1, ANSWER: 3, AUTHORITY: 0, ADDITIONAL: 0

;; QUESTION SECTION:
;yahoo.com.                    IN      MX

;; ANSWER SECTION:
yahoo.com.            5        IN      MX      1 mta7.am0.yahoodns.net.
yahoo.com.            5        IN      MX      1 mta5.am0.yahoodns.net.
yahoo.com.            5        IN      MX      1 mta6.am0.yahoodns.net.

;; Query time: 63 msec
;; SERVER: 192.168.182.2#53(192.168.182.2)
;; WHEN: Thu Jan 17 20:08:12 2019
;; MSG SIZE  rcvd: 106

m1:Desktop honp$ ▊
```

In the preceding screenshot, we see that Yahoo! uses three mail servers. Each server has a preference of `1`. This means that mail can be delivered to any one of them with no special preference. If mail delivery fails to the first chosen server, then another server should be chosen at random to retry.

Programmatically getting the MX record in a cross-platform manner can be difficult. Please see `Chapter 5`, *Hostname Resolution and DNS*, where this topic was covered in some depth. The SMTP client we develop in this present chapter assumes that the mail server is known in advance.

Now that we know which server to connect to, let's consider the SMTP protocol itself in more detail.

SMTP dialog

SMTP is a text-based TCP protocol that works on port 25. SMTP works in a lock-step, one-at-a-time dialog, with the client sending commands and the server sending responses for each command.

In a typical session, the dialog goes as follows:

1. The client first establishes a connection to the SMTP server.
2. The server initiates with a greeting. This greeting indicates that the server is ready to receive commands.
3. The client then issues its own greeting.
4. The server responds.
5. The client sends a command indicating who the mail is from.
6. The server responds to indicate that the sender is accepted.
7. The client issues another command, which specifies the mail's recipient.
8. The server responds indicating the recipient is accepted.
9. The client then issues a DATA command.
10. The server responds asking the client to proceed.
11. The client transfers the email.

The protocol is very simple. In the following example SMTP session, mail.example.net is the client, and the server is mail.example.com (C and S indicate whether the client or server is sending, respectively):

```
S: 220 mail.example.com SMTP server ready
C: HELO mail.example.net
S: 250 Hello mail.example.net [192.0.2.67]
C: MAIL FROM:<alice@example.net>
S: 250 OK
C: RCPT TO:<bob@example.com>
S: 250 Accepted
C: DATA
S: 354 Enter message, ending with "." on a line by itself
C: Subject: Re: The Cake
C: Date: Fri, 03 May 2019 02:31:20 +0000
C:
C: Do NOT forget to bring the cake!
C: .
S: 250 OK
C: QUIT
S: 221 closing connection
```

Everything the server sends is in reply to the client's commands, except for the first line. The first line is simply in response to the client connecting.

You may notice that each of the client's commands start with a four-letter word. Each one of the server's responses starts with a three-digit code.

The common client commands we use are as follows:

- HELO is used for the client to identify itself to the server.
- MAIL is used to specify who is sending the mail.
- RCPT is used to specify a recipient.
- DATA is used to initiate the transfer of the actual email. This email should include both headers and a body.
- QUIT is used to end the session.

The server response codes used in a successful email transfer are the following:

- 220: The service is ready
- 250: The requested command was accepted and completed successfully
- 354: Start sending the message
- 221: The connection is closing

Error codes vary between providers, but they are generally in the 500 range.

SMTP servers can also send replies spanning multiple lines. In this case, the very last line begins with the response code followed by a space. All preceding lines begin with the response code followed by a dash. The following example illustrates a multiline response after attempting to deliver to a mailbox that does not exist:

```
C: MAIL FROM:<alice@example.net>
S: 250 OK
C: RCPT TO:<not-a-real-user@example.com>
S: 550-The account you tried to deliver to does not
S: 550-exist. Please double-check the recipient's
S: 550 address for typos and try again.
```

Note that some servers validate that the recipient address is valid before replying to the RCPT command, but many servers only validate the recipient address after the client sends the email using the DATA command.

Although that explains the basics of the protocol used to send mail, we still must consider the format of the email itself. This is covered next.

The format of an email

If we make an analogy to physical mail, the SMTP commands `MAIL FROM` and `RCPT TO` address the envelope. Those commands give the SMTP server information on how the mail is to be delivered. In this analogy, the `DATA` command would be the letter inside the envelope. As it's common to address a physical letter inside an envelope, it's also common to repeat the delivery information in the email, even though it was already sent to the SMTP server through the `MAIL` and `RCPT` commands.

A simple email may look like the following:

```
From: Alice Doe <alice@example.net>
To: Bob Doe <bob@example.com>
Subject: Re: The Cake
Date: Fri, 03 May 2019 02:31:20 +0000

Hi Bob,

Do NOT forget to bring the cake!

Best,
Alice
```

The entire email is transmitted to an SMTP server following the `DATA` command. A single period on an otherwise blank line is transmitted to indicate the end of the email. If the email contains any line beginning with a period, the SMTP client should replace it with two consecutive periods. This prevents the client from indicating the email is over prematurely. The SMTP server knows that any line beginning with two periods should be replaced with a single period.

The email itself can be divided into two parts—the header and the body. The two parts are delineated by the first blank line.

The header part consists of various headers that indicate properties of the email. `From`, `To`, `Subject`, and `Date` are the most common headers.

The body part of the email is simply the message being sent.

With a basic understanding of the email format, we are now ready to begin writing a simple C program to send emails.

A simple SMTP client program

With a basic understanding of both SMTP and the email format, we are ready to program a simple email client. Our client takes as inputs: the destination email server, the recipient's address, the sender's address, the email subject line, and the email body text.

Our program begins by including the necessary headers with the following statements:

```
/*smtp_send.c*/

#include "chap08.h"
#include <ctype.h>
#include <stdarg.h>
```

We also define the following two constants to make buffer allocation and checking easier:

```
/*smtp_send.c continued*/

#define MAXINPUT 512
#define MAXRESPONSE 1024
```

Our program needs to prompt the user for input several times. This is required to get the email server's hostname, the recipient's address, and so on. C provides the gets() function for this purpose but gets() is deprecated in the latest C standard. Therefore, we implement our own function.

The following function, get_input(), prompts for user input:

```
/*smtp_send.c continued*/

void get_input(const char *prompt, char *buffer)
{
    printf("%s", prompt);

    buffer[0] = 0;
    fgets(buffer, MAXINPUT, stdin);
    const int read = strlen(buffer);
    if (read > 0)
        buffer[read-1] = 0;
}
```

The get_input() function uses fgets() to read from stdin. The buffer passed to get_input() is assumed to be MAXINPUT bytes, which we defined at the top of the file.

The `fgets()` function does not remove a newline character from the received input; therefore, we overwrite the last character inputted with a terminating null character.

It is also very helpful to have a function that can send formatted strings directly over the network. We implement a function called `send_format()` for this purpose. It takes a socket, a formatting string, and the additional arguments to send. You can think of `send_format()` as being very similar to `printf()`. The difference is that `send_format()` delivers the formatted text over the network instead of printing to the screen.

The code for `send_format()` is as follows:

```
/*smtp_send.c continued*/

void send_format(SOCKET server, const char *text, ...) {
    char buffer[1024];
    va_list args;
    va_start(args, text);
    vsprintf(buffer, text, args);
    va_end(args);

    send(server, buffer, strlen(buffer), 0);

    printf("C: %s", buffer);
}
```

The preceding code works by first reserving a buffer. `vsprintf()` is then used to format the text into that buffer. It is up to the caller to ensure that the formatted output doesn't exceed the reserved buffer space. We are assuming for this program that the user is trusted, but in a production program, you would want to add checks to prevent a buffer overflow here.

After the output text is formatted into `buffer`, it is sent using `send()`. We also print the sent text to the screen. A `C:` is printed preceding it to indicate that the text was sent by us, the client.

One of the trickier parts of our SMTP client is parsing the SMTP server responses. This is important because the SMTP client must not issue a second command until a response is received for the first command. If the SMTP client sends a new command before the server is ready, then the server will likely terminate the connection.

Recall that each SMTP response starts with a three-digit code. We want to parse out this code to check for errors. Each SMTP response is usually followed by text that we ignore. SMTP responses are typically only one line long, but they can sometimes span multiple lines. In this case, each line up to the penultimate line contains a dash character, –, directly following the three-digit response code.

To illustrate how multiline responses work, consider the following two responses as equivalent:

```
/*response 1*/

250 Message received!

/*response 2*/

250-Message
250 received!
```

It is important that our program recognizes multiline responses; it must not mistakenly treat a single multiline response as separate responses.

We implement a function called `parse_response()` for this purpose. It takes in a null-terminated response string and returns the parsed response code. If no code is found or the response isn't complete, then `0` is returned instead. The code for this function is as follows:

```
/*smtp_send.c continued*/

int parse_response(const char *response) {
    const char *k = response;
    if (!k[0] || !k[1] || !k[2]) return 0;
    for (; k[3]; ++k) {
        if (k == response || k[-1] == '\n') {
            if (isdigit(k[0]) && isdigit(k[1]) && isdigit(k[2])) {
                if (k[3] != '-') {
                    if (strstr(k, "\r\n")) {
                        return strtol(k, 0, 10);
                    }
                }
            }
        }
    }
    return 0;
}
```

The `parse_response()` function begins by checking for a null terminator in the first three characters of the response. If a null is found there, then the function can return immediately because `response` isn't long enough to constitute a valid SMTP response.

It then loops through the `response` input string. The loop goes until a null-terminating character is found three characters out. Each loop, `isdigit()` is used to see whether the current character and the next two characters are all digits. If so, the fourth character, `k[3]`, is checked. If `k[3]` is a dash, then the response continues onto the next line. However, if `k[3]` isn't a dash, then `k[0]` represents the beginning of the last line of the SMTP response. In this case, the code checks if the line ending has been received; `strstr()` is used for this purpose. It the line ending was received, the code uses `strtol()` to convert the response code to an integer.

If the code loops through `response()` without returning, then 0 is returned, and the client needs to wait for more input from the SMTP server.

With `parse_response()` out of the way, it is useful to have a function that waits until a particular response code is received over the network. We implement a function called `wait_on_response()` for this purpose, which begins as follows:

```
/*smtp_send.c continued*/

void wait_on_response(SOCKET server, int expecting) {
    char response[MAXRESPONSE+1];
    char *p = response;
    char *end = response + MAXRESPONSE;

    int code = 0;
```

In the preceding code, a `response` buffer variable is reserved for storing the SMTP server's response. A pointer, `p`, is set to the beginning of this buffer; `p` will be incremented to point to the end of the received data, but it starts at `response` since no data has been received yet. An `end` pointer variable is set to the end of the buffer, which is useful to ensure we do not attempt to write past the buffer end.

Finally, we set `code = 0` to indicate that no response code has been received yet.

The `wait_on_response()` function then continues with a loop as follows:

```
/*smtp_send.c continued*/

do {
    int bytes_received = recv(server, p, end - p, 0);
    if (bytes_received < 1) {
        fprintf(stderr, "Connection dropped.\n");
        exit(1);
    }

    p += bytes_received;
```

```
        *p = 0;

        if (p == end) {
            fprintf(stderr, "Server response too large:\n");
            fprintf(stderr, "%s", response);
            exit(1);
        }

        code = parse_response(response);

    } while (code == 0);
```

The beginning of the preceding loop uses recv() to receive data from the SMTP server. The received data is written at point p in the response array. We are careful to use end to make sure received data isn't written past the end of response.

After recv() returns, p is incremented to the end of the received data, and a null terminating character is set. A check for p == end ensures that we haven't written to the end of the response buffer.

Our function from earlier, parse_response(), is used to check whether a full SMTP response has been received. If so, then code is set to that response. If not, then code is equal to 0, and the loop continues to receive additional data.

After the loop terminates, the wait_on_response() function checks that the received SMTP response code is as expected. If so, the received data is printed to the screen, and the function returns. The code for this is as follows:

```
    /*smtp_send.c continued*/

    if (code != expecting) {
        fprintf(stderr, "Error from server:\n");
        fprintf(stderr, "%s", response);
        exit(1);
    }

    printf("S: %s", response);
}
```

That concludes the wait_on_response() function. This function proves very useful, and it is needed after every command sent to the SMTP server.

We also define a function called connect_to_host(), which attempts to open a TCP connection to a given hostname and port number. This function is extremely similar to the code we've used in the previous chapters.

First `getaddrinfo()` is used to resolve the hostname and `getnameinfo()` is then used to print the server IP address. The following code achieves those two purposes:

```
/*smtp_send.c continued*/

SOCKET connect_to_host(const char *hostname, const char *port) {
    printf("Configuring remote address...\n");
    struct addrinfo hints;
    memset(&hints, 0, sizeof(hints));
    hints.ai_socktype = SOCK_STREAM;
    struct addrinfo *peer_address;
    if (getaddrinfo(hostname, port, &hints, &peer_address)) {
        fprintf(stderr, "getaddrinfo() failed. (%d)\n", GETSOCKETERRNO());
        exit(1);
    }

    printf("Remote address is: ");
    char address_buffer[100];
    char service_buffer[100];
    getnameinfo(peer_address->ai_addr, peer_address->ai_addrlen,
            address_buffer, sizeof(address_buffer),
            service_buffer, sizeof(service_buffer),
            NI_NUMERICHOST);
    printf("%s %s\n", address_buffer, service_buffer);
```

A socket is then created with `socket()`, as shown in the following code:

```
/*smtp_send.c continued*/

    printf("Creating socket...\n");
    SOCKET server;
    server = socket(peer_address->ai_family,
            peer_address->ai_socktype, peer_address->ai_protocol);
    if (!ISVALIDSOCKET(server)) {
        fprintf(stderr, "socket() failed. (%d)\n", GETSOCKETERRNO());
        exit(1);
    }
```

Once the socket has been created, `connect()` is used to establish the connection. The following code shows the use of `connect()` and the end of the `connect_to_host()` function:

```
/*smtp_send.c continued*/

    printf("Connecting...\n");
    if (connect(server,
                peer_address->ai_addr, peer_address->ai_addrlen)) {
        fprintf(stderr, "connect() failed. (%d)\n", GETSOCKETERRNO());
```

```
        exit(1);
    }
    freeaddrinfo(peer_address);

    printf("Connected.\n\n");

    return server;
}
```

Don't forget to call `freeaddrinfo()` to free the memory allocated for the server address, as shown by the preceding code.

Finally, with those helper functions out of the way, we can begin on `main()`. The following code defines `main()` and initializes Winsock if required:

```
/*smtp_send.c continued*/

int main() {

#if defined(_WIN32)
    WSADATA d;
    if (WSAStartup(MAKEWORD(2, 2), &d)) {
        fprintf(stderr, "Failed to initialize.\n");
        return 1;
    }
#endif
```

See `Chapter 2`, *Getting to Grips with Socket APIs*, for more information about initializing Winsock and establishing connections.

Our program can proceed by prompting the user for an SMTP hostname. This hostname is stored in `hostname`, and our `connect_to_host()` function is used to open a connection. The following code shows this:

```
/*smtp_send.c continued*/

char hostname[MAXINPUT];
get_input("mail server: ", hostname);

printf("Connecting to host: %s:25\n", hostname);

SOCKET server = connect_to_host(hostname, "25");
```

After the connection is established, our SMTP client must not issue any commands until the server responds with a `220` code. We use `wait_on_response()` to wait for this with the following code:

```
/*smtp_send.c continued*/

    wait_on_response(server, 220);
```

Once the server is ready to receive commands, we must issue the `HELO` command. The following code sends the `HELO` command and waits for a `250` response code:

```
/*smtp_send.c continued*/

    send_format(server, "HELO HONPWC\r\n");
    wait_on_response(server, 250);
```

`HELO` should be followed by the SMTP client's hostname; however, since we are probably running this client from a development machine, it's likely we don't have a hostname setup. For this reason, we simply send `HONPWC`, although any arbitrary string can be used. If you are running this client from a server, then you should change the `HONPWC` string to a domain that points to your server.

Also, note the line ending used in the preceding code. The line ending used by SMTP is a carriage return character followed by a line feed character. In C, this is represented by `"\r\n"`.

Our program then prompts the user for the sending and receiving addresses and issues the appropriate SMTP commands. This is done using `get_input()` to prompt the user, `send_format()` to issue the SMTP commands, and `wait_on_response()` to receive the SMTP server's response:

```
/*smtp_send.c continued*/

    char sender[MAXINPUT];
    get_input("from: ", sender);
    send_format(server, "MAIL FROM:<%s>\r\n", sender);
    wait_on_response(server, 250);

    char recipient[MAXINPUT];
    get_input("to: ", recipient);
    send_format(server, "RCPT TO:<%s>\r\n", recipient);
    wait_on_response(server, 250);
```

After the sender and receiver are specified, the next step in the SMTP is to issue the DATA command. The DATA command instructs the server to listen for the actual email. It is issued by the following code:

```
/*smtp_send.c continued*/

    send_format(server, "DATA\r\n");
    wait_on_response(server, 354);
```

Our client program then prompts the user for an email subject line. After the subject line is specified, it can send the email headers: From, To, and Subject. The following code does this:

```
/*smtp_send.c continued*/

    char subject[MAXINPUT];
    get_input("subject: ", subject);

    send_format(server, "From:<%s>\r\n", sender);
    send_format(server, "To:<%s>\r\n", recipient);
    send_format(server, "Subject:%s\r\n", subject);
```

It is also useful to add a date header. Emails use a special format for dates. We can make use of the strftime() function to format the date properly. The following code formats the date into the proper email header:

```
/*smtp_send.c continued*/

    time_t timer;
    time(&timer);

    struct tm *timeinfo;
    timeinfo = gmtime(&timer);

    char date[128];
    strftime(date, 128, "%a, %d %b %Y %H:%M:%S +0000", timeinfo);

    send_format(server, "Date:%s\r\n", date);
```

In the preceding code, the time() function is used to get the current date and time, and gmtime() is used to convert it into a timeinfo struct. Then, strftime() is called to format the data and time into a temporary buffer, date. This formatted string is then transmitted to the SMTP server as an email header.

After the email headers are sent, the email body is delineated by a blank line. The following code sends this blank line:

```
/*smtp_send.c continued*/

    send_format(server, "\r\n");
```

We can then prompt the user for the body of the email using `get_input()`. The body is transmitted one line at a time. When the user finishes their email, they should enter a single period on a line by itself. This indicates both to our client and the SMTP server that the email is finished.

The following code sends user input to the server until a single period is inputted:

```
/*smtp_send.c continued*/

    printf("Enter your email text, end with \".\" on a line by itself.\n");

    while (1) {
        char body[MAXINPUT];
        get_input("> ", body);
        send_format(server, "%s\r\n", body);
        if (strcmp(body, ".") == 0) {
            break;
        }
    }
```

If the mail was accepted by the SMTP server, it sends a `250` response code. Our client then issues the `QUIT` command and checks for a `221` response code. The `221` response code indicates that the connection is terminating as shown in the following code:

```
/*smtp_send.c continued*/

    wait_on_response(server, 250);

    send_format(server, "QUIT\r\n");
    wait_on_response(server, 221);
```

Our SMTP client concludes by closing the socket, cleaning up Winsock (if required), and exiting as shown here:

```
/*smtp_send.c continued*/

    printf("\nClosing socket...\n");
    CLOSESOCKET(server);

#if defined(_WIN32)
```

```
    WSACleanup();
#endif

    printf("Finished.\n");
    return 0;
}
```

That concludes smtp_send.c.

You can compile and run smtp_send.c on Windows with MinGW using the following:

```
gcc smtp_send.c -o smtp_send.exe -lws2_32
smtp_send.exe
```

On Linux or macOS, compiling and running smtp_send.c is done by the following:

```
gcc smtp_send.c -o smtp_send
./smtp_send
```

The following screenshot shows the sending of a simple email using smtp_send.c:

If you're doing a lot of testing, you may find it tedious to type in the email each time. In that case, you can automate it by putting your text in a file and using the `cat` utility to read it into `smtp_send`. For example, you may have the `email.txt` file as follows:

```
/*email.txt*/

mail-server.example.net
bob@example.com
alice@example.net
Re: The Cake
Hi Alice,

What about the cake then?

Bob
.
```

With the program input stored in `email.txt`, you can send an email using the following command:

```
cat email.txt | ./smtp_send
```

Hopefully, you can send some test emails with `smtp_send`. There are, however, a few obstacles you may run into. Your ISP may block outgoing emails from your connection, and many email servers do not accept mail from residential IP address blocks. See the *Spam-blocking pitfalls* section later in the chapter for more information.

Although `smtp_send` is useful for sending simple text-based messages, you may be wondering how to add formatting to your email. Perhaps you want to send files as attachments. The next section addresses these issues.

Enhanced emails

The emails we've been looking at so far have been only simple text. Modern email usage often demands fancier formatted emails.

We can control the content type of an email using the `Content-Type` header. This is very similar to the content type header used by HTTP, which we covered in `Chapter 7`, *Building a Simple Web Server*.

If the content type header is missing, a content type of `text/plain` is assumed by default. Therefore, the `Content-Type` header in the following email is redundant:

```
From: Alice Doe <alice@example.net>
To: Bob Doe <bob@example.com>
Subject: Re: The Cake
Date: Fri, 03 May 2019 02:31:20 +0000
Content-Type: text/plain

Hi Bob,

Do NOT forget to bring the cake!

Best,
Alice
```

If you want formatting support in your email, which is common today, you should use a `text/html` content type. In the following email, HTML is used to add emphasis:

```
From: Alice Doe <alice@example.net>
To: Bob Doe <bob@example.com>
Subject: Re: The Cake
Date: Fri, 03 May 2019 02:31:20 +0000
Content-Type: text/html

Hi Bob,<br>
<br>
Do <strong>NOT</strong> forget to bring the cake!<br>
<br>
Best,<br>
Alice<br>
```

Not all email clients support HTML emails. For this reason, it may be useful to encode your message as both plaintext and HTML. The following email uses this technique:

```
From: Alice Doe <alice@example.net>
To: Bob Doe <bob@example.com>
Subject: Re: The Cake
Date: Fri, 03 May 2019 02:31:20 +0000
MIME-Version: 1.0
Content-Type: multipart/alternative; boundary="SEPARATOR"

This is a message with multiple parts in MIME format.
--SEPARATOR
Content-Type: text/plain

Hi Bob,
```

```
Do NOT forget to bring the cake!

Best,
Alice
--SEPARATOR
Content-Type: text/html

Hi Bob,<br>
<br>
Do <strong>NOT</strong> forget to bring the cake!<br>
<br>
Best,<br>
Alice<br>
--SEPARATOR--
```

The preceding email example uses two headers to indicate that it's a multipart message. The first one, MIME-Version: 1.0, indicates which version of **Multipurpose Internet Mail Extensions** (**MIME**) we're using. MIME is used for all emails that aren't simply plaintext.

The second header, Content-Type: multipart/alternative; boundary="SEPARATOR", specifies that we're sending a multipart message. It also specifies a special boundary character sequence that delineates the parts of the email. In our example, SEPARATOR is used as the boundary. It is important that the boundary not appear in the email text or attachments. In practice, boundary specifiers are often long randomly generated strings.

Once the boundary has been set, each part of the email begins with --SEPARATOR on a line by itself. The email ends with --SEPARATOR--. Note that each part of the message gets its own header section, specific to only that part. These header sections are used to specify the content type for each part.

It is also often useful to attach files to email, which we will cover now.

Email file attachments

If a multipart email is being sent, a part can be designated as an attachment using the Content-Disposition header. See the following example:

```
From: Alice Doe <alice@example.net>
To: Bob Doe <bob@example.com>
Subject: Re: The Cake
Date: Fri, 03 May 2019 02:31:20 +0000
MIME-Version: 1.0
```

```
Content-Type: multipart/alternative; boundary="SEPARATOR"

This is a message with multiple parts in MIME format.
--SEPARATOR
Content-Type: text/plain

Hi Bob,

Please see the attached text file.

Best,
Alice
--SEPARATOR
Content-Disposition: attachment; filename=my_file.txt;
  modification-date="Fri, 03 May 2019 02:26:51 +0000";
Content-Type: application/octet-stream
Content-Transfer-Encoding: base64

VGhpcyBpcyBhIHNpbXBsZSB0ZXh0IG1lc3NhZ2Uu
--SEPARATOR--
```

The preceding example includes a file called `my_file.txt`. SMTP is a purely text-based protocol; therefore, any attachments that may include binary data need to be encoded into a text format. Base64 encoding is commonly used for this purpose. In this example, the header, `Content-Transfer-Encoding: base64`, specifies that we are going to use Base64 encoding.

The content of `my_file.txt` is `This is a simple text message`. That sentence encodes to Base64 as `VGhpcyBpcyBhIHNpbXBsZSB0ZXh0IG1lc3NhZ2Uu` as seen in the preceding code.

Spam-blocking pitfalls

It can be much harder to send emails today than it was in the past. Spam has become a major problem, and every provider is taking actions to curb it. Unfortunately, many of these actions can also make it much more difficult to send legitimate emails.

Many residential ISPs don't allow outgoing connections on port 25. If your residential provider blocks port 25, then you won't be able to establish an SMTP connection. In this case, you may consider renting a virtual private server to run this chapter's code.

Even if your ISP does allow outgoing connections on port 25, many SMTP servers won't accept mail from a residential IP address. Of the servers that do, many will send those emails straight into a spam folder.

For example, if you attempt to deliver an email to Gmail, you may get a response similar to the following:

```
550-5.7.1 [192.0.2.67] The IP you're using to send mail is not authorized
550-5.7.1 to send email directly to our servers. Please use the SMTP
550-5.7.1 relay at your service provider instead. Learn more at
550 5.7.1  https://support.google.com/mail/?p=NotAuthorizedError
```

Another spam-blocking measure that may trip you up is **Sender Policy Framework** (**SPF**). SPF works by listing which servers can send mail for a given domain. If a sending server isn't on the SPF list, then receiving SMTP servers reject their mail.

DomainKeys Identified Mail (**DKIM**) is a measure to authenticate email using digital signatures. Many popular email providers are more likely to treat non-DKIM mail as spam. DKIM signing is very complicated and out of scope for this book.

Domain-based Message Authentication, Reporting, and Conformance (**DMARC**) is a technique used for domains to publish whether SPF and/or DKIM is required of mail originating from them, among other things.

Most commercial email servers use SPF, DKIM, and DMARC. If you're sending email without these, your email will likely be treated as spam. If you're sending email in opposition to these, your email will either be rejected outright or labeled as spam.

Finally, many popular providers assign a reputation to sending domains and servers. This puts potential emails senders in a catch-22 situation. Email can't be delivered without building a reputation, but a reputation can't be built without successfully delivering lots of emails. As spam continues to be a major problem, we may soon come to a time where only big-name, trusted SMTP servers can operate with each other. Let's hope it doesn't come to that!

If your program needs to send email reliably, you should likely consider using an email service provider. One option is to allow an SMTP relay to do the final delivery. A potentially easier option is to use a mail sending service that operates an HTTP API.

Summary

In this chapter, we looked at how email is delivered over the internet. SMTP, the protocol responsible for email delivery, was studied in some depth. We then constructed a simple program to send short emails using SMTP.

We looked at the email format too. We saw how MIME could be used to send multipart emails with file attachments.

We also saw how sending emails over the modern internet is full of pitfalls. Many of these stems from attempts to block spam. Techniques used by providers, such as blocking residential IP addresses, SPF, DKIM, DMARC, and IP address reputation monitoring, may make it difficult for our simple program to deliver email reliably.

In the next chapter, `Chapter 9`, *Loading Secure Web Pages with HTTPS and OpenSSL*, we look at secure web connections using HTTPS.

Questions

Try these questions to test your knowledge from this chapter:

1. What port does SMTP operate on?
2. How do you determine which SMTP server receives mail for a given domain?
3. How do you determine which SMTP server sends mail for a given provider?
4. Why won't an SMTP server relay mail without authentication?
5. How are binary files sent as email attachments when SMTP is a text-based protocol?

The answers can be found in `Appendix A`, *Answers to Questions*.

Further reading

For more information about SMTP and email formats, please refer to the following links:

- **RFC 821**: *Simple Mail Transfer Protocol* (`https://tools.ietf.org/html/rfc821`)
- **RFC 2822**: *Internet Message Format* (`https://tools.ietf.org/html/rfc2822`)

3
Section 3 - Understanding Encrypted Protocols and OpenSSL

In this section, the reader will learn about the secure, encrypted protocols that are common on the modern web.

The following chapters are in this section:

9
Loading Secure Web Pages with HTTPS and OpenSSL

In this chapter, we'll learn how to initiate secure connections to web servers using **Hypertext Transfer Protocol Secure** (**HTTPS**). HTTPS provides several benefits over HTTP. HTTPS gives an authentication method to identify servers and detect impostors. It also protects the privacy of all transmitted data and prevents interceptors from tampering with or forging transmitted data.

In HTTPS, communication is secured using **Transport Layer Security** (**TLS**). In this chapter, we'll learn how to use the popular OpenSSL library to provide TLS functionality.

The following topics are covered in this chapter:

- Background information about HTTPS
- Types of encryption ciphers
- How servers are authenticated
- Basic OpenSSL usage
- Creating a simple HTTPS client

Technical requirements

The example programs from this chapter can be compiled using any modern C compiler. We recommend MinGW for Windows and GCC for Linux and macOS. You also need to have the OpenSSL library installed. See `Appendices B`, *Setting Up Your C Compiler on Windows*, `Appendices C`, *Setting Up Your C Compiler on Linux*, and `Appendices D`, *Setting Up Your C Compiler on macOS*, for compiler setup and OpenSSL installation.

The code for this book can be found at `https://github.com/codeplea/Hands-On-Network-Programming-with-C`.

From the command line, you can download the code for this chapter with the following command:

```
git clone https://github.com/codeplea/Hands-On-Network-Programming-with-C
cd Hands-On-Network-Programming-with-C/chap09
```

Each example program in this chapter runs on Windows, Linux, and macOS. While compiling on Windows, each example program requires linking with the Winsock library. This can be accomplished by passing the `-lws2_32` option to `gcc`.

Each example also needs to be linked against the OpenSSL libraries, `libssl.a` and `libcrypto.a`. This is accomplished by passing `-lssl -lcrypto` to `gcc`.

We provide the exact commands needed to compile each example as it is introduced.

All of the example programs in this chapter require the same header files and C macros that we developed in `Chapter 2`, *Getting to Grips with Socket APIs*. For brevity, we put these statements in a separate header file, `chap09.h`, which we can include in each program. For an explanation of these statements, please refer to `Chapter 2`, *Getting to Grips with Socket APIs*.

The content of `chap09.h` begins by including the necessary networking header files. The code for this follows:

```
#if defined(_WIN32)
#ifndef _WIN32_WINNT
#define _WIN32_WINNT 0x0600
#endif
#include <winsock2.h>
#include <ws2tcpip.h>
#pragma comment(lib, "ws2_32.lib")

#else
#include <sys/types.h>
#include <sys/socket.h>
#include <netinet/in.h>
#include <arpa/inet.h>
#include <netdb.h>
#include <unistd.h>
#include <errno.h>

#endif
```

We also define some macros to assist with writing portable code like so:

```
/*chap09.h continued*/

#if defined(_WIN32)
#define ISVALIDSOCKET(s) ((s) != INVALID_SOCKET)
#define CLOSESOCKET(s) closesocket(s)
#define GETSOCKETERRNO() (WSAGetLastError())

#else
#define ISVALIDSOCKET(s) ((s) >= 0)
#define CLOSESOCKET(s) close(s)
#define SOCKET int
#define GETSOCKETERRNO() (errno)
#endif
```

Finally, `chap09.h` includes some additional headers, including the headers for the OpenSSL library. This is shown in the following code:

```
/*chap09.h continued*/

#include <stdio.h>
#include <string.h>
#include <stdlib.h>
#include <time.h>

#include <openssl/crypto.h>
#include <openssl/x509.h>
#include <openssl/pem.h>
#include <openssl/ssl.h>
#include <openssl/err.h>
```

HTTPS overview

HTTPS provides security to HTTP. We covered HTTP in `Chapter 6`, *Building a Simple Web Client*. HTTPS secures HTTP by using TLS over TCP on port 443. TLS is a protocol that can provide security to any TCP connection.

TLS is the successor to **Secure Socket Layer** (**SSL**), an earlier protocol also used by HTTPS. TLS and SSL are compatible, and most of the information in this chapter also applies to SSL. Generally, establishing an HTTPS connection involves the client and server negotiating which protocol to use. The ideal outcome is that the client and server agree on the most secure, mutually supported protocol and cipher.

When we talk about protocol security, we are generally looking for the following three things:

- **Authentication**: We need a way to prevent impostors from posing as legitimate communication partners. TLS provides peer authentication methods for this reason.
- **Encryption**: TLS uses encryption to obfuscate transmitted data. This prevents an eavesdropper from correctly interpreting intercepted data.
- **Integrity**: TLS also ensures that received data has not been tampered with or otherwise forged.

HTTP is most commonly used to transmit web pages. The text on a web page is first encoded as **Hypertext Markup Language** (**HTML**). **HTML** provides formatting, layout, and styling to web pages. **HTTP** is then used to transmit the **HTML**, and **HTTP** itself is transmitted over a **TCP** connection.

Visually, an **HTTP** session is encapsulated like the following:

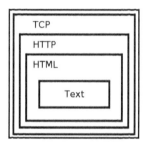

TLS works inside TCP to provide a secure communication channel. HTTPS is then basically the same as the **HTTP** protocol, but it is sent inside a **TLS** channel.

Visually, HTTPS is encapsulated in the following manner:

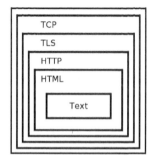

Of course, the same abstraction still applies if **HTTP** is used to transmit an image, video, or other data instead of **HTML**.

Do keep in mind that these abstractions are accurate at the conceptual level, but some details transcend layers. For example, some HTTPS headers are used to refer to security parameters of how **TLS** should be applied. In general, though, it is reasonable to think of **TLS** as securing the **TCP** connection used by HTTPS.

Although **TLS** is most commonly used for HTTPS security, **TLS** is also used to secure many other **TCP**-based protocols. The email protocol, SMTP, which we covered in `Chapter 8`, *Making Your Program Send Email*, is also commonly secured by **TLS**.

Before going into further detail about using **TLS**, it is useful to understand some necessary background information on encryption.

Encryption basics

Encryption is a method of encoding data so that only authorized parties can access it. Encryption does not prevent interception or interference, but it denies the original data to a would-be attacker.

Encryption algorithms are called **ciphers**. An encryption cipher takes unencrypted data as input, referred to as **plaintext**. The cipher produces encrypted data, called **ciphertext**, as its output. The act of converting plaintext into ciphertext is called encryption, and the act of reversing it back is called **decryption**.

Modern ciphers use keys to control the encryption and decryption of data. Keys are typically relatively short, pseudo-random data sequences. Ciphertext encrypted with a given key cannot be decrypted without the proper key.

Broadly, there are two categories of ciphers—symmetric and asymmetric. A symmetric cipher uses the same key for both encryption and decryption, while an asymmetric cipher uses two different keys.

Symmetric ciphers

The following diagram illustrates a symmetric cipher:

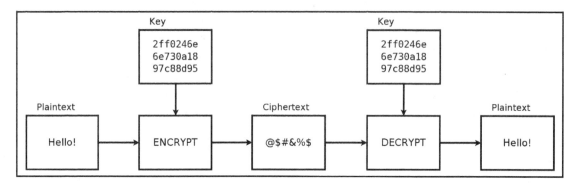

In the preceding diagram, the plaintext **Hello!** is encrypted using a symmetric cipher. A secret key is used with the cipher to produce a ciphertext. This ciphertext can then be transmitted over an insecure channel, and eavesdroppers cannot decipher it without knowledge of the secret key. The privileged receiver of the ciphertext uses the decryption algorithm and secret key to convert it back into plaintext.

Some symmetric ciphers in general use (not just for TLS) are the following:

- **American Encryption Standard (AES)**, also known as **Rijndael**
- Camellia
- **Data Encryption Standard (DES)**
- Triple DES
- **International Data Encryption Algorithm (IDEA)**
- QUAD
- RC4
- Salsa20, Chacha20
- **Tiny Encryption Algorithm (TEA)**

One issue with symmetric encryption is that the same key must be known to both the sender and receiver. Generating and transmitting this key securely poses a problem. How can the key be sent between parties if they don't already have a secure communication channel? If they do already have a secure communication channel, why is encryption needed in the first place?

Key exchange algorithms attempt to address these problems. Key exchange algorithms work by allowing both communicating parties to generate the same secret key. In general, the parties first agree on a public, non-secret key. Then, each party generates its own secret key and combines it with the public key. These combined keys are exchanged. Each party then adds its own secret to the combined keys to arrive at a combined, secret key. This combined, secret key is then known to both parties, but not derivable by an eavesdropper.

The most common key exchange algorithm in use today is the **Diffie-Hellman key exchange algorithm**.

While key exchange algorithms are resistant against eavesdroppers, they are not resilient to interception. In the case of interception, an attacker could stand in the middle of a key exchange, while posing as each corresponding party. This is called a **man-in-the-middle attack**.

Asymmetric ciphers can be used to address some of these problems.

Asymmetric ciphers

Asymmetric encryption, also known as **public-key encryption**, attempts to solve the key exchange and authentication problems of symmetric encryption. With asymmetric encryption, two keys are used. One key can encrypt data, while the other key can decrypt it. These keys are generated together and related mathematically. However, deriving one key from the other after the fact is not possible.

The following diagram shows an asymmetric cipher:

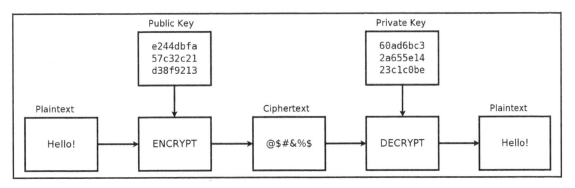

Establishing a secure communication channel with asymmetric encryption is easier. Each party can generate its own asymmetric encryption keys. The encryption key can then be transmitted without the worry of interception, while the decryption key is kept private. In this scheme, these keys are referred to as the **Public Key** and the **Private Key**, respectively.

The **Rivest-Shamir-Adleman** (**RSA**) cipher is one of the first public-key ciphers and is widely used today. Newer **elliptic-curve cryptography** (**ECC**) algorithms promise greater efficiency and are quickly gaining market share.

Asymmetric ciphers are also used to implement digital signatures. A digital signature is used to verify the authenticity of data. Digital signatures are created by using a private key to generate a signature for a document. The public key can then be used to verify that the document was signed by the holder of the private key.

The following diagram illustrates a digital signing and verifying process:

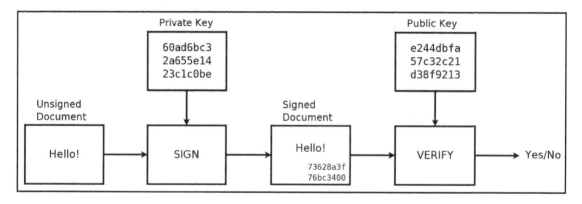

TLS uses a combination of these methods to achieve security.

How TLS uses ciphers

Digital signatures are essential in TLS; they are used to authenticate servers. Without digital signatures, a TLS client wouldn't be able to differentiate between an authentic server and an impostor.

TLS can also use digital signatures to authenticate the client, although this is much less common in practice. Most web applications either don't care about client authentication, or use other simpler methods, such as passwords.

In theory, asymmetric encryption could be used to protect an entire communication channel. However, in practice, modern asymmetric ciphers are inefficient and only able to protect small amounts of data. For this reason, symmetric ciphers are preferred whenever possible. TLS uses asymmetric ciphers only to authenticate the server. TLS uses a key exchange algorithm and symmetric ciphers to protect the actual communication.

Vulnerabilities are found for encryption algorithms all the time. It is, therefore, imperative that TLS connections are able to select the best algorithms that are mutually supported by both parties. This is done using **cipher suites**. A cipher suite is a list of algorithms, generally a **key exchange algorithm**, a **bulk encryption algorithm**, and a **message authentication algorithm** (**MAC**).

When a TLS connection is first established, the TLS client sends a list of preferred **cipher suites** to the server. The TLS server will select one of these cipher suites to be used for the connection. If the server doesn't support any of the cipher suites given by the client, then a secure TLS connection cannot be established.

With some background information about security out of the way, we can discuss TLS in more detail.

The TLS protocol

After a TCP connection is established, the TLS handshake is initiated by the client. The client sends a number of specifications to the server, including which versions of SSL/TLS it is running, which cipher suites it supports, and which compression methods it would like to use.

The server selects the highest mutually supported version of SSL/TLS to use. It also chooses a cipher suite and compression method from the choices given by the client.

If the client and server do not support any cipher suite in common, then no TLS connection can be established. This is not uncommon when using very old browsers with newer servers.

After the basic setup is done, the server sends the client its certificate. This is used by the client to verify that it's connected to a legitimate server. We'll discuss more on certificates in the next section.

Once the client has verified that the server really is who it claims to be, a key exchange is initiated. After key exchange completes, both the client and server have a shared secret key. All further communication is encrypted using this key and their chosen symmetric cipher.

Certificates are used to verify server identities with digital signatures. Let's explore how they work next.

Certificates

Each HTTPS server uses one or more certificates to verify their identity. This certificate must either be trusted by the client itself or trusted by a third party that the client trusts. In common usages, such as web browsers, it's not really possible to list all trusted certificates. For this reason, it's most common to validate certificates by verifying that a trusted third party trusts them. This trust is proven using digital signatures.

For example, a popular digital certificate authority is DigiCert Inc. Suppose that you trust DigiCert Inc. and have stored a certificate from them locally; you then connect to a website, example.com. You may not trust example.com, because you haven't seen their certificate before. However, example.com shows you that their certificate has been digitally signed by the DigiCert Inc. certificate you do trust. Therefore, you trust example.com website's certificate too.

In practice, certificate chains can be several layers deep. As long as you can verify digital signatures back to a certificate you trust, you are able to validate the whole chain.

This method is the common one used by HTTPS to authenticate servers. It does have some issues; namely, you must entirely trust the certificate authority. This is because certificate authorities could theoretically issue a certificate to an impostor, in which case you would be forced to trust the impostor. Certificate authorities are careful to avoid this, as it would ruin their reputation.

The most popular certificate authorities at the time of this writing are the following:

- IdenTrust
- Comodo
- DigiCert
- GoDaddy
- GlobalSign

The five preceding certificate authorities are responsible for over 90% of the HTTPS certificates found on the web.

It is also possible to self-sign a certificate. In this case, no certificate authority is used. In these cases, a client needs to somehow reliably obtain and verify a copy of your certificate before it can be trusted.

Certificates are usually matched to domain names, but it's also possible for them to identify other information, such as company names, address, and so on.

In the next chapter, `Chapter 10`, *Implementing a Secure Web Server*, certificates are covered in more detail.

It's common today for one server to host many different domains, and each domain requires its own certificate. Let's now consider how these servers know which certificate to send.

Server name identification

Many servers host multiple domains. Certificates are tied to domains; therefore, TLS must provide a method for the client to specify which domain it's connecting to. You may recall that the HTTP *Host* header servers this purpose. The problem is that the TLS connection should be established before the HTTP data is sent. Therefore, the server must decide which certificate to transmit before the HTTP *Host* header is received.

This is accomplished using **Server Name Indication** (**SNI**). SNI is a technique that, when used by TLS, requires the client to indicate to the server which domain it is attempting to connect to. The server can then find a matching certificate to use for the TLS connection.

SNI is relatively new, and older browsers and servers do not support it. Before SNI was popular, servers had two choices—they could either host only one domain per IP address, or they could send certificates for all hosted domains for each connection.

It should be noted that SNI involves sending the unencrypted domain name over the network. This means an eavesdropper can see which host the client is connecting to, even though they wouldn't know which resources the client is requesting from that host. Newer protocols, such as **encrypted server name identification** (**ESNI**), address this problem but are not widely deployed yet.

With a basic understanding of the TLS protocol, we're ready to look at the most popular library that implements it—OpenSSL.

OpenSSL

OpenSSL is a widely used open source library that provides SSL and TLS services to applications. We use it in this chapter for secure connections required by HTTPS.

OpenSSL can be challenging to install. Refer to Appendices B, *Setting Up Your C Compiler on Windows*, Appendices C, *Setting Up Your C Compiler on Linux*, and Appendices D, *Setting Up Your C Compiler on macOS*, for more information.

You can check whether you have the OpenSSL command-line tools installed by running the following command:

```
openssl version
```

The following screenshot shows this on Ubuntu Linux:

You'll also need to ensure that you have the OpenSSL library installed. The following program can be used to test this. If it compiles and runs successfully, then you do have the OpenSSL library installed and working:

```
/*openssl_version.c*/

#include <openssl/ssl.h>

int main(int argc, char *argv[]) {
    printf("OpenSSL version: %s\n", OpenSSL_version(SSLEAY_VERSION));
    return 0;
}
```

If you're using an older version of OpenSSL, you may need to replace the OpenSSL_version() function call with SSLeay_version() instead. However, a better solution is to just upgrade to a newer OpenSSL version.

The preceding `openssl_version.c` program is compiled on macOS and Linux using the following command:

```
gcc openssl_version.c -o openssl_version -lcrypto
./openssl_version
```

The following screenshot shows compiling and running `openssl_version.c`:

On Windows, `openssl_version` can be compiled using MinGW and the following commands:

```
gcc openssl_version.c -o openssl_version.exe -lcrypto
openssl_version.exe
```

Once the OpenSSL library is installed and usable, we are ready to begin using encrypted sockets.

Encrypted sockets with OpenSSL

The TLS provided by OpenSSL can be applied to any TCP socket.

Before using OpenSSL in your program, it is important to initialize it. The following code does this:

```
SSL_library_init();
OpenSSL_add_all_algorithms();
SSL_load_error_strings();
```

In the preceding code, the call to `SSL_library_init()` is required to initialize the OpenSSL library. The second call (to `OpenSSL_add_all_algorithms()`) causes OpenSSL to load all available algorithms. Alternately, you could load only the algorithms you know are needed. For our purposes, it is easy to load them all. The third call, `SSL_load_error_strings()`, causes OpenSSL to load error strings. This call isn't strictly needed, but it is handy to have easily readable error messages when something goes wrong.

Once OpenSSL is initialized, we are ready to create an SSL context. This is done by calling the `SSL_CTX_new()` function, which returns an `SSL_CTX` object. You can think of this object as a sort of factory for SSL/TLS connections. It holds the initial settings that you want to use for your connections. Most programs only need to create one `SSL_CTX` object, and they can reuse it for all their connections.

The following code creates the SSL context:

```
SSL_CTX *ctx = SSL_CTX_new(TLS_client_method());
if (!ctx) {
    fprintf(stderr, "SSL_CTX_new() failed.\n");
    return 1;
}
```

The `SSL_CTX_new()` function takes one argument. We use `TLS_client_method()`, which indicates that we want general-purpose, version-flexible TLS methods available. Our client automatically negotiates the best mutually supported algorithm with the server upon connecting.

If you're using an older version of OpenSSL, you may need to replace `TLS_client_method()` with `TLSv1_2_client_method()` in the preceding code. However, a much better solution is to upgrade to a newer OpenSSL version.

To secure a TCP connection, you must first have a TCP connection. This TCP connection should be established in the normal way. The following pseudo-code shows this:

```
getaddrinfo(hostname, port, hints, address);
socket = socket(address, type, protocol);
connect(socket, address, type);
```

For more information on setting up a TCP connection, refer to Chapter 3, *An In-Depth Overview of TCP Connections*.

Once `connect()` has returned successfully, and a TCP connection is established, you can use the following code to initiate a TLS connection:

```
SSL *ssl = SSL_new(ctx);
if (!ctx) {
    fprintf(stderr, "SSL_new() failed.\n");
    return 1;
}

if (!SSL_set_tlsext_host_name(ssl, hostname)) {
    fprintf(stderr, "SSL_set_tlsext_host_name() failed.\n");
    ERR_print_errors_fp(stderr);
    return 1;
}

SSL_set_fd(ssl, socket);
if (SSL_connect(ssl) == -1) {
    fprintf(stderr, "SSL_connect() failed.\n");
    ERR_print_errors_fp(stderr);
    return 1;
}
```

In the preceding code, `SSL_new()` is used to create an `SSL` object. This object is used to track the new SSL/TLS connection.

We then use `SSL_set_tlsext_host_name()` to set the domain for the server we are trying to connect to. This allows OpenSSL to use SNI. This call is optional, but without it, the server does not know which certificate to send in the case that it hosts more than one site.

Finally, we call `SSL_set_fd()` and `SSL_connect()` to initiate the new TLS/SSL connection on our existing TCP socket.

It is possible to see which cipher the TLS connection is using. The following code shows this:

```
printf ("SSL/TLS using %s\n", SSL_get_cipher(ssl));
```

Once the TLS connection is established, data can be sent and received using `SSL_write()` and `SSL_read()`, respectively. These functions are used in a nearly identical manner as the standard socket `send()` and `recv()` functions.

The following example shows transmitting a simple message over a TLS connection:

```
char *data = "Hello World!";
int bytes_sent = SSL_write(ssl, data, strlen(data));
```

Receiving data, done with `SSL_read()`, is shown in the following example:

```
char read[4096];
int bytes_received = SSL_read(ssl, read, 4096);
printf("Received: %.*s\n", bytes_received, read);
```

When the connection is finished, it's important to free the used resources by calling `SSL_shutdown()` and `SSL_free(ssl)`. This is shown in the following code:

```
SSL_shutdown(ssl);
CLOSESOCKET(socket);
SSL_free(ssl);
```

When you're done with an SSL context, you should also call `SSL_CTX_free()`. In our case, it looks like this:

```
SSL_CTX_free(ctx);
```

If your program requires authentication of the connected peer, it is important to look at the certificates sent during the TLS initialization. Let's consider that next.

Certificates

Once the TLS connection is established, we can use the `SSL_get_peer_certificate()` function to get the server's certificate. It's also easy to print the certificate subject and issuer, as shown in the following code:

```
X509 *cert = SSL_get_peer_certificate(ssl);
if (!cert) {
    fprintf(stderr, "SSL_get_peer_certificate() failed.\n");
    return 1;
}

char *tmp;
if (tmp = X509_NAME_oneline(X509_get_subject_name(cert), 0, 0)) {
    printf("subject: %s\n", tmp);
    OPENSSL_free(tmp);
}

if (tmp = X509_NAME_oneline(X509_get_issuer_name(cert), 0, 0)) {
    printf("issuer: %s\n", tmp);
    OPENSSL_free(tmp);
}

X509_free(cert);
```

OpenSSL automatically verifies the certificate during the TLS/SSL handshake. You can get the verification results using the `SSL_get_verify_result()` function. Its usage is shown in the following code:

```
long vp = SSL_get_verify_result(ssl);
if (vp == X509_V_OK) {
    printf("Certificates verified successfully.\n");
} else {
    printf("Could not verify certificates: %ld\n", vp);
}
```

If `SSL_get_verify_result()` returns `X509_V_OK`, then the certificate chain was verified by OpenSSL and the connection can be trusted. If `SSL_get_verify_result()` does not return `X509_V_OK`, then HTTPS authentication has failed, and the connection should be abandoned.

In order for OpenSSL to successfully verify the certificate, we must tell it which certificate authorities we trust. This can be done by using the `SSL_CTX_load_verify_locations()` function. It must be passed the filename that stores all of the trusted root certificates. Assuming your trusted certificates are in `trusted.pem`, the following code sets this up:

```
if (!SSL_CTX_load_verify_locations(ctx, "trusted.pem", 0)) {
    fprintf(stderr, "SSL_CTX_load_verify_locations() failed.\n");
    ERR_print_errors_fp(stderr);
    return 1;
}
```

Deciding which root certificates to trust isn't easy. Each operating system provides a list of trusted certificates, but there is no general, easy way to import these lists. Using the operating system's default list is also not always appropriate for every application. For these reasons, certificate verification has been omitted from the examples in this chapter. However, it is absolutely critical that it be implemented appropriately in order for your HTTPS connections to be secure.

In addition to validating the certificate's signatures, it is also important to validate that the certificate is actually valid for the particular server you're connected with! Newer versions of OpenSSL provide functions to help with this, but with older versions of OpenSSL, you're on your own. Consult the OpenSSL documentation for more information.

We've now covered enough background information about TLS and OpenSSL that we're ready to tackle a concrete example program.

A simple HTTPS client

To bring the concepts of this chapter together, we build a simple HTTPS client. This client can connect to a given HTTPS web server and request the root document /.

Our program begins by including the needed chapter header, defining `main()`, and initializing Winsock as shown here:

```
/*https_simple.c*/

#include "chap09.h"

int main(int argc, char *argv[]) {

#if defined(_WIN32)
    WSADATA d;
    if (WSAStartup(MAKEWORD(2, 2), &d)) {
        fprintf(stderr, "Failed to initialize.\n");
        return 1;
    }
#endif
```

We then initialize the OpenSSL library with the following code:

```
/*https_simple.c continued*/

SSL_library_init();
OpenSSL_add_all_algorithms();
SSL_load_error_strings();
```

The `SSL_load_error_strings()` function call is optional, but it's very useful if we run into problems.

We can also create an OpenSSL context. This is done by calling `SSL_CTX_new()` as shown by the following code:

```
/*https_simple.c continued*/

SSL_CTX *ctx = SSL_CTX_new(TLS_client_method());
if (!ctx) {
    fprintf(stderr, "SSL_CTX_new() failed.\n");
    return 1;
}
```

If we were going to do certificate verification, this would be a good place to include the `SSL_CTX_load_verify_locations()` function call, as explained in the *Certificates* section of this chapter. We're omitting certification verification in this example for simplicity, but it is important to include it in real-world applications.

Our program then checks that a hostname and a port number was passed in on the command line like so:

```
/*https_simple.c continued*/

    if (argc < 3) {
        fprintf(stderr, "usage: https_simple hostname port\n");
        return 1;
    }

    char *hostname = argv[1];
    char *port = argv[2];
```

The standard HTTPS port number is `443`. Our program lets the user specify any port number, which can be useful for testing.

We then configure the remote address for the socket connection. This code uses the same technique that we've been using since `Chapter 3`, *An In-Depth Overview of TCP connections*. The code for this is as follows:

```
/*https_simple.c continued*/

    printf("Configuring remote address...\n");
    struct addrinfo hints;
    memset(&hints, 0, sizeof(hints));
    hints.ai_socktype = SOCK_STREAM;
    struct addrinfo *peer_address;
    if (getaddrinfo(hostname, port, &hints, &peer_address)) {
        fprintf(stderr, "getaddrinfo() failed. (%d)\n", GETSOCKETERRNO());
        exit(1);
    }

    printf("Remote address is: ");
    char address_buffer[100];
    char service_buffer[100];
    getnameinfo(peer_address->ai_addr, peer_address->ai_addrlen,
            address_buffer, sizeof(address_buffer),
            service_buffer, sizeof(service_buffer),
            NI_NUMERICHOST);
    printf("%s %s\n", address_buffer, service_buffer);
```

We continue to create the socket using a call to `socket()`, and we connect it using the `connect()` function as follows:

```
/*https_simple.c continued*/

    printf("Creating socket...\n");
    SOCKET server;
    server = socket(peer_address->ai_family,
            peer_address->ai_socktype, peer_address->ai_protocol);
    if (!ISVALIDSOCKET(server)) {
        fprintf(stderr, "socket() failed. (%d)\n", GETSOCKETERRNO());
        exit(1);
    }

    printf("Connecting...\n");
    if (connect(server,
                peer_address->ai_addr, peer_address->ai_addrlen)) {
        fprintf(stderr, "connect() failed. (%d)\n", GETSOCKETERRNO());
        exit(1);
    }
    freeaddrinfo(peer_address);

    printf("Connected.\n\n");
```

At this point, a TCP connection has been established. If we didn't need encryption, we could communicate over it directly. However, we are going to use OpenSSL to initiate a TLS/SSL connection over our TCP connection. The following code creates a new SSL object, sets the hostname for SNI, and initiates the TLS/SSL handshake:

```
/*https_simple.c continued*/

    SSL *ssl = SSL_new(ctx);
    if (!ctx) {
        fprintf(stderr, "SSL_new() failed.\n");
        return 1;
    }

    if (!SSL_set_tlsext_host_name(ssl, hostname)) {
        fprintf(stderr, "SSL_set_tlsext_host_name() failed.\n");
        ERR_print_errors_fp(stderr);
        return 1;
    }

    SSL_set_fd(ssl, server);
    if (SSL_connect(ssl) == -1) {
        fprintf(stderr, "SSL_connect() failed.\n");
        ERR_print_errors_fp(stderr);
```

```
        return 1;
    }
```

The preceding code is explained in the *Encrypted Sockets with OpenSSL* section earlier in this chapter.

The call to `SSL_set_tlsext_host_name()` is optional, but useful if you may be connecting to a server that hosts multiple domains. Without this call, the server wouldn't know which certificates are relevant to this connection.

It is sometimes useful to know which cipher suite the client and server agreed upon. We can print the selected cipher suite with the following:

```
/*https_simple.c continued*/

    printf("SSL/TLS using %s\n", SSL_get_cipher(ssl));
```

It is also useful to see the server's certificate. The following code prints the server's certificate:

```
/*https_simple.c continued*/

    X509 *cert = SSL_get_peer_certificate(ssl);
    if (!cert) {
        fprintf(stderr, "SSL_get_peer_certificate() failed.\n");
        return 1;
    }

    char *tmp;
    if ((tmp = X509_NAME_oneline(X509_get_subject_name(cert), 0, 0))) {
        printf("subject: %s\n", tmp);
        OPENSSL_free(tmp);
    }

    if ((tmp = X509_NAME_oneline(X509_get_issuer_name(cert), 0, 0))) {
        printf("issuer: %s\n", tmp);
        OPENSSL_free(tmp);
    }

    X509_free(cert);
```

The certificate *subject* should match the domain we're connecting to. The issuer should be a certificate authority that we trust. Note that the preceding code does *not* validate the certificate. Refer to the *Certificates* section in this chapter for more information.

We can then send our HTTPS request. This request is the same as if we were using plain HTTP. We begin by formatting the request into a buffer and then sending it over the encrypted connection using `SSL_write()`. The following code shows this:

```
/*https_simple.c continued*/

    char buffer[2048];

    sprintf(buffer, "GET / HTTP/1.1\r\n");
    sprintf(buffer + strlen(buffer), "Host: %s:%s\r\n", hostname, port);
    sprintf(buffer + strlen(buffer), "Connection: close\r\n");
    sprintf(buffer + strlen(buffer), "User-Agent: https_simple\r\n");
    sprintf(buffer + strlen(buffer), "\r\n");

    SSL_write(ssl, buffer, strlen(buffer));
    printf("Sent Headers:\n%s", buffer);
```

For more information about the HTTP protocol, please refer back to Chapter 6, *Building a Simple Web Client.*

Our client now simply waits for data from the server until the connection is closed. This is accomplished by using `SSL_read()` in a loop. The following code receives the HTTPS response:

```
/*https_simple.c continued*/

    while(1) {
        int bytes_received = SSL_read(ssl, buffer, sizeof(buffer));
            if (bytes_received < 1) {
                printf("\nConnection closed by peer.\n");
                break;
            }

            printf("Received (%d bytes): '%.*s'\n",
                    bytes_received, bytes_received, buffer);
    }
```

The preceding code also prints any data received over the HTTPS connection. Note that it does not parse the received HTTP data, which would be more complicated. See the `https_get.c` program in this chapter's code repository for a more advanced program that does parse the HTTP response.

Our simple client is almost done. We only have to shut down the TLS/SSL connection, close the socket, and clean up. This is done by the following code:

```
/*https_simple.c continued*/

    printf("\nClosing socket...\n");
    SSL_shutdown(ssl);
    CLOSESOCKET(server);
    SSL_free(ssl);
    SSL_CTX_free(ctx);

#if defined(_WIN32)
    WSACleanup();
#endif

    printf("Finished.\n");
    return 0;
}
```

That concludes `https_simple.c`.

You should be able to compile and run it on Windows using MinGW and the following commands:

```
gcc https_simple.c -o https_simple.exe -lssl -lcrypto -lws2_32
https_simple example.org 443
```

If you're using certain older versions of OpenSSL, you may also need an additional linker option—`-lgdi32`.

Compiling and executing on macOS and Linux can be done using the following:

```
gcc https_simple.c -o https_simple -lssl -lcrypto
./https_simple example.org 443
```

If you run into linker errors, you should check that your OpenSSL library is properly installed. You may find it helpful to attempt compiling the `openssl_version.c` program first.

The following screenshot shows successfully compiling `https_simple.c` and using it to connect to `example.org`:

```
honp@ubby18: ~/Desktop
$ gcc https_simple.c -o https_simple -lssl -lcrypto -Wall
$ ./https_simple example.org 443
Configuring remote address...
Remote address is: 93.184.216.34 https
Creating socket...
Connecting...
Connected.

SSL/TLS using ECDHE-RSA-AES128-GCM-SHA256
subject: /C=US/ST=California/L=Los Angeles/O=Internet Corporation for
Assigned Names and Numbers/OU=Technology/CN=www.example.org
issuer: /C=US/O=DigiCert Inc/CN=DigiCert SHA2 Secure Server CA
Sent Headers:
GET / HTTP/1.1
Host: example.org:443
Connection: close
User-Agent: https_simple

Received (341 bytes): 'HTTP/1.1 200 OK
Cache-Control: max-age=604800
Content-Type: text/html; charset=UTF-8
Date: Thu, 24 Jan 2019 23:50:51 GMT
Etag: "1541025663+ident"
Expires: Thu, 31 Jan 2019 23:50:51 GMT
Last-Modified: Fri, 09 Aug 2013 23:54:35 GMT
Server: ECS (ord/4CDA)
Vary: Accept-Encoding
X-Cache: HIT
Content-Length: 1270
Connection: close


Received (1270 bytes): '<!doctype html>
<html>
<head>
    <title>Example Domain</title>
...

Connection closed by peer.

Closing socket...
Finished.
$
```

The `https_simple` program should serve as an elementary example of the techniques of connecting as an HTTPS client. These same techniques can be applied to any TCP connection.

It's easy to apply these techniques to some of the programs developed earlier in the book, such as `tcp_client` and `web_get`.

It is also worth mentioning that, while TLS works only with TCP connections, **Datagram Transport Layer Security** (**DTLS**) aims to provide many of the same guarantees for **User Datagram Protocol** (**UDP**) datagrams. OpenSSL provides support for DTLS too.

Other examples

A few other examples are included in this chapter's repository. They are the following:

- `tls_client.c`: This is `tcp_client.c` from `Chapter 3`, *An In-Depth Overview of TCP Connections*, but it has been modified to make TLS connections.
- `https_get.c`: This is the `web_get.c` program from `Chapter 6`, *Building a Simple Web Client*, but it has been modified for HTTPS. You can think of it as the extended version of `https_simple.c`.
- `tls_get_cert.c`: This is like `https_simple.c`, but it simply prints the connected server's certificate and exits.

Keep in mind that none of the examples in this chapter perform certification verification. This is an important step that must be added before using these programs in the field.

Summary

In this chapter, we learned about the features that HTTPS provides over HTTP, such as authentication and encryption. We saw that HTTPS is really just HTTP over a TLS connection and that TLS can be applied to any TCP connection.

We also learned about basic encryption concepts. We saw how asymmetric ciphers use two keys, and how this allows for digital signatures. The very basics of certificates were covered, and we explored some of the difficulties with verifying them.

Finally, we worked through a concrete example that established a TLS connection to an HTTPS server.

This chapter was all about HTTPS clients, but in the next chapter, we focus on how HTTPS servers work.

Questions

Try these questions to test your knowledge from this chapter:

1. What port does HTTPS typically operate on?
2. How many keys does symmetric encryption use?
3. How many keys does asymmetric encryption use?
4. Does TLS use symmetric or asymmetric encryption?
5. What is the difference between SSL and TLS?
6. What purpose do certificates fill?

Answers are in `Appendix A`, *Answers to Questions*.

Further reading

For more information about HTTPS and OpenSSL, please refer to the following:

- OpenSSL documentation (`https://www.openssl.org/docs/`)
- **RFC 5246**: *The Transport Layer Security (TLS) Protocol* (`https://tools.ietf.org/html/rfc5246`)

10
Implementing a Secure Web Server

In this chapter, we will build a simple HTTPS server program. This serves as the counterpart to the HTTPS client we worked on in the previous chapter.

HTTPS is powered by **Transport Layer Security** (**TLS**). HTTPS servers, unlike HTTPS clients, are expected to identify themselves with certificates. We'll cover how to listen for HTTPS connections, provide certificates, and send an HTTP response over TLS.

The following topics are covered in this chapter:

- HTTPS overview
- HTTPS certificates
- HTTPS server setup with OpenSSL
- Accepting HTTPS connections
- Common problems
- OpenSSL alternatives
- Direct TLS termination alternatives

Technical requirements

This chapter continues where `Chapter 9`, *Loading Secure Web Pages with HTTPS and OpenSSL*, left off. This chapter continues to use the OpenSSL library. It is imperative that you have the OpenSSL library installed and that you know the basics of programming with OpenSSL. Refer to `Chapter 9`, *Loading Secure Web Pages with HTTPS and OpenSSL*, for basic information about OpenSSL.

The example programs from this chapter can be compiled using any modern C compiler. We recommend **MinGW** for Windows and **GCC** for Linux and macOS. You also need to have the OpenSSL library installed. See `Appendix B`, *Setting Up Your C Compiler on Windows*; `Appendix C`, *Setting Up Your C Compiler on Linux*; and `Appendix D`, *Setting Up Your C Compiler on macOS*, for compiler setup and OpenSSL installation.

The code for this book can be found at `https://github.com/codeplea/Hands-On-Network-Programming-with-C`.

From the command line, you can download the code for this chapter by using the following command:

```
git clone https://github.com/codeplea/Hands-On-Network-Programming-with-C
cd Hands-On-Network-Programming-with-C/chap10
```

Each example program in this chapter runs on Windows, Linux, and macOS. When compiling on Windows, each example program requires linking with the **Winsock** library. This is accomplished by passing the `-lws2_32` option to `gcc`.

Each example also needs to be linked against the OpenSSL libraries, `libssl.a` and `libcrypto.a`. This is accomplished by passing `-lssl -lcrypto` to GCC.

We provide the exact commands needed to compile each example as it is introduced.

All of the example programs in this chapter require the same header files and C macros that we developed in `Chapter 2`, *Getting to Grips with Socket APIs*. For brevity, we put these statements in a separate header file, `chap10.h`, which we can include in each program. For an explanation of these statements, please refer to `Chapter 2`, *Getting to Grips with Socket APIs*.

The content of `chap10.h` begins by including the required networking header files. The code for this is as follows:

```
/*chap10.h*/

#if defined(_WIN32)
#ifndef _WIN32_WINNT
#define _WIN32_WINNT 0x0600
#endif
#include <winsock2.h>
#include <ws2tcpip.h>
#pragma comment(lib, "ws2_32.lib")

#else
#include <sys/types.h>
```

```
#include <sys/socket.h>
#include <netinet/in.h>
#include <arpa/inet.h>
#include <netdb.h>
#include <unistd.h>
#include <errno.h>

#endif
```

We also define some macros to assist with writing portable code:

```
/*chap10.h continued*/

#if defined(_WIN32)
#define ISVALIDSOCKET(s) ((s) != INVALID_SOCKET)
#define CLOSESOCKET(s) closesocket(s)
#define GETSOCKETERRNO() (WSAGetLastError())

#else
#define ISVALIDSOCKET(s) ((s) >= 0)
#define CLOSESOCKET(s) close(s)
#define SOCKET int
#define GETSOCKETERRNO() (errno)
#endif
```

Finally, chap10.h includes the additional headers required by this chapter's programs:

```
/*chap10.h continued*/

#include <stdio.h>
#include <string.h>
#include <stdlib.h>
#include <time.h>

#include <openssl/crypto.h>
#include <openssl/x509.h>
#include <openssl/pem.h>
#include <openssl/ssl.h>
#include <openssl/err.h>
```

HTTPS and OpenSSL summary

We begin with a quick review of the HTTPS protocol, as covered in Chapter 9, *Loading Secure Web Pages with HTTPS and OpenSSL*. However, we do recommend that you work through Chapter 9, *Loading Secure Web Pages with HTTPS and OpenSSL*, before beginning this chapter.

HTTPS uses TLS to add security to HTTP. You will recall from Chapter 6, *Building a Simple Web Client*, and Chapter 7, *Building a Simple Web Server*, that HTTP is a text-based protocol that works over TCP on port 80. The TLS protocol can be used to add security to any TCP-based protocol. Specifically, TLS is used to provide security for HTTPS. So in a nutshell, HTTPS is simply HTTP with TLS. The default HTTPS port is 443.

OpenSSL is a popular open source library that provides functionality for TLS/SSL and HTTPS. We use it in this book to provide the methods needed to implement HTTPS clients and servers.

Generally, HTTPS connections are first made using TCP sockets. Once the TCP connection is established, OpenSSL is used to negotiate a TLS connection over the open TCP connection. From that point forward, OpenSSL functions are used to send and receive data over the TLS connection.

An important part of communication security is being able to trust that the connection is to the intended party. No amount of data encryption helps if you have connected to an impostor. TLS uses certificates to prevent against connecting to impostors and man-in-the-middle attacks.

We now need to understand certificates in more detail before we can proceed with our HTTPS server.

Certificates

Certificates are an important part of the TLS protocol. Although certificates can be used on both the client and server side, HTTPS generally only uses server certificates. These certificates identify to the client that they are connected to a trusted server.

Without certificates, a client wouldn't be able to tell whether it was connected to the intended server or an impostor server.

Certificates work using a **chain-of-trust** model. Each HTTPS client has a few certificate authorities that they explicitly trust, and certificate authorities offer services where they digitally sign certificates. This service is usually done for a small fee, and usually only after some simple verification of the requester.

When an HTTPS client sees a certificate signed by an authority it trusts, then it also trusts that certificate. Indeed, these chains of trust can run deep. For example, most certificate authorities also allow resellers. In these cases, the certificate authority signs an intermediate certificate to be used by the reseller. The reseller then uses this intermediate certificate to sign new certificates. Clients trust the intermediate certificates because they are signed by trusted certificate authorities, and clients trust the certificates signed by resellers because they trust the intermediate certificates.

Certificate authorities commonly offer two types of validation. **Domain validation** is where a signed certificate is issued after simply verifying that the certificate recipient can be reached at the given domain. This is usually done by having the certificate requester temporarily modify a DNS record, or reply to an email sent to their *whois* contact.

Let's Encrypt is a relatively new certificate issuer that issues certificates for free. They do this using an automated model. **Domain validation** is done by having the certificate requester serve a small file over HTTP or HTTPS.

Domain validation is the most common type of validation. An HTTPS server using domain validation assures an HTTPS client that they are connected to the domain that they think they are. It implies that their connection wasn't silently hijacked or otherwise intercepted.

Certificate authorities also offer **Extended Validation** (**EV**) certificates. EV certificates are only issued after the authority verifies the recipient's identity. This is usually done using public records and a phone call.

For public-facing HTTPS applications, it is important that you obtain a certificate from a recognized certificate authority. However, this can sometimes be tedious, and it is often much more convenient to obtain a **self-signed** certificate for development and testing purposes. Let's do that now.

Self-signed certificates with OpenSSL

Certificates signed by recognized authorities are essential to establish the chain of trust needed for public websites. However, it is much easier to obtain a self-signed certificate for testing or development.

It is also acceptable to use a self-signed certificate for certain private applications where the client can be deployed with a copy of the certificate, and trust only that certificate. This is called **certificate pinning**. Indeed, when used properly, certificate pinning can be more secure than using a certificate authority. However, it is not appropriate for public-facing websites.

We require a certificate to test our HTTPS server. We use a self-signed certificate because they are the easiest to obtain. The downside to this method is that web browsers won't trust our server. We can get around this by clicking through a few warnings in the web browser.

OpenSSL provides tools to make self-signing certificates very easy.

The basic command to self-sign a certificate is as follows:

```
openssl req -x509 -newkey rsa:2048 -nodes -sha256 -keyout key.pem \
-out cert.pem -days 365
```

OpenSSL asks questions about what to put on the certificate, including the subject, your name, company, location, and so on. You can use the defaults on all of these as this doesn't matter for our testing purposes.

The preceding command places the new certificate in cert.pem and the key for it in key.pem. Our HTTPS server needs both files. cert.pem is the certificate that gets sent to the connected client, and key.pem provides our server with the encryption key that proves that it owns the certificate. Keeping this key secret is imperative.

Here is a screenshot showing the generation of a new self-signed certificate:

```
                        honp@ubby18: ~/Desktop
$ openssl req -x509 -newkey rsa:2048 -nodes -sha256 -keyout key.pem -out cert.pem
-days 36500
Generating a 2048 bit RSA private key
.............+++
..............................................................................
.+++
writing new private key to 'key.pem'
-----
You are about to be asked to enter information that will be incorporated
into your certificate request.
What you are about to enter is what is called a Distinguished Name or a DN.
There are quite a few fields but you can leave some blank
For some fields there will be a default value,
If you enter '.', the field will be left blank.
-----
Country Name (2 letter code) [AU]:
State or Province Name (full name) [Some-State]:
Locality Name (eg, city) []:
Organization Name (eg, company) [Internet Widgits Pty Ltd]:
Organizational Unit Name (eg, section) []:
Common Name (e.g. server FQDN or YOUR name) []:
Email Address []:
$
```

You can also use OpenSSL to view a certificate. The following command does this:

```
openssl x509 -text -noout -in cert.pem
```

If you're on Windows using MSYS, you may get garbled line endings from the previous command. If so, try using unix2dos to fix it, as shown by the following command:

```
openssl x509 -text -noout -in cert.pem | unix2dos
```

Here is what a typical self-signed certificate looks like:

```
                           honp@ubby18: ~/Desktop                        ⊖ ⊡ ⊗
$ openssl x509 -text -noout -in cert.pem
Certificate:
    Data:
        Version: 3 (0x2)
        Serial Number:
            d5:e8:c8:77:6a:18:c6:80
    Signature Algorithm: sha256WithRSAEncryption
        Issuer: C = AU, ST = Some-State, O = Internet Widgits Pty Ltd
        Validity
            Not Before: Feb  7 21:10:06 2019 GMT
            Not After : Jan 14 21:10:06 2119 GMT
        Subject: C = AU, ST = Some-State, O = Internet Widgits Pty Ltd
        Subject Public Key Info:
            Public Key Algorithm: rsaEncryption
                Public-Key: (2048 bit)
                Modulus:
                    00:de:21:8c:e1:d1:37:a1:1b:57:96:f4:bc:1d:8f:
                    da:c3:4b:a7:2e:85:f7:2d:17:f7:87:ac:a0:67:cb:
                    c3:08:c1:f5:e3:67:fd:8b:46:8a:91:23:a2:8e:e9:
                    f0:c4:3e:94:ad:d7:3e:dc:7f:55:ce:b8:ed:85:f5:
                    b1:9a:0b:ad:71:fe:0e:22:82:a8:bc:9f:60:f2:94:
                    a2:68:32:b2:32:58:e5:2d:60:75:62:24:b7:ac:cc:
                    cc:19:18:1d:54:1c:d0:4c:b3:91:3d:44:de:fb:6d:
```

Now that we have a usable certificate, we are ready to begin our HTTPS server programming.

HTTPS server with OpenSSL

Let's go over some basics of using the OpenSSL library in server applications before beginning a concrete example.

Before OpenSSL can be used, it must be initialized. The following code initializes the OpenSSL library, loads the requisite encryption algorithms, and loads useful error strings:

```
SSL_library_init();
OpenSSL_add_all_algorithms();
SSL_load_error_strings();
```

Refer to the previous Chapter 9, *Loading Secure Web Pages with HTTPS and OpenSSL,* for more information.

Our server also needs to create an SSL context object. This object works as a sort of factory from which we can create TLS/SSL connections.

The following code creates the `SSL_CTX` object:

```
SSL_CTX *ctx = SSL_CTX_new(TLS_server_method());
if (!ctx) {
    fprintf(stderr, "SSL_CTX_new() failed.\n");
    return 1;
}
```

If you're using an older version of OpenSSL, you may need to replace `TLS_server_method()` with `TLSv1_2_server_method()` in the preceding code. However, a better solution is to upgrade to a newer OpenSSL version.

After the `SSL_CTX` object is created, we can set it to use our self-signed certificate and key. The following code does this:

```
if (!SSL_CTX_use_certificate_file(ctx, "cert.pem" , SSL_FILETYPE_PEM)
|| !SSL_CTX_use_PrivateKey_file(ctx, "key.pem", SSL_FILETYPE_PEM)) {
    fprintf(stderr, "SSL_CTX_use_certificate_file() failed.\n");
    ERR_print_errors_fp(stderr);
    return 1;
}
```

That concludes the minimal OpenSSL setup needed for an HTTPS server.

The server should then listen for incoming TCP connections. This was covered in detail in `Chapter 3`, *An In-Depth Overview of TCP Connections.*

After a new TCP connection is established, we use the socket returned by `accept()` to create our TLS/SSL socket.

First, a new `SSL` object is created using our SSL context from earlier. The following code demonstrates this:

```
SSL *ssl = SSL_new(ctx);
if (!ctx) {
    fprintf(stderr, "SSL_new() failed.\n");
    return 1;
}
```

The `SSL` object is then linked to our open TCP socket using `SSL_set_fd()`. The `SSL_accept()` function is called to establish the TLS/SSL connection. The following code demonstrates this:

```
SSL_set_fd(ssl, socket_client);
if (SSL_accept(ssl) <= 0) {
    fprintf(stderr, "SSL_accept() failed.\n");
    ERR_print_errors_fp(stderr);
```

```
        return 1;
    }

    printf ("SSL connection using %s\n", SSL_get_cipher(ssl));
```

You may notice that this code is very similar to the HTTPS client code from Chapter 9, *Loading Secure Web Pages with HTTPS and OpenSSL*. The only real differences are in the setup of the SSL context object.

Once the TLS connection is established, data can be sent and received using SSL_write() and SSL_read(). These functions replace the send() and recv() functions used with TCP sockets.

When the connection is finished, it is important to free resources, as shown by the following code:

```
SSL_shutdown(ssl);
CLOSESOCKET(socket_client);
SSL_free(ssl);
```

When your program is finished accepting new connections, you should also free the SSL context object. The following code shows this:

```
SSL_CTX_free(ctx);
```

With an understanding of the basics out of the way, let's solidify our knowledge by implementing a simple example program.

Time server example

In this chapter, we develop a simple time server that displays the time to an HTTPS client. This program is an adaptation of time_server.c from Chapter 2, *Getting to Grips with Socket APIs*, which served the time over plain HTTP. Our program begins by including the chapter header, defining main(), and initializing Winsock on Windows. The code for this is as follows:

```
/*tls_time_server.c*/

#include "chap10.h"

int main() {

#if defined(_WIN32)
    WSADATA d;
    if (WSAStartup(MAKEWORD(2, 2), &d)) {
```

```
        fprintf(stderr, "Failed to initialize.\n");
        return 1;
    }
#endif
```

The OpenSSL library is then initialized with the following code:

```
/*tls_time_server.c continued*/
    SSL_library_init();
    OpenSSL_add_all_algorithms();
    SSL_load_error_strings();
```

An SSL context object must be created for our server. This is done using a call to
SSL_CTX_new(). The following code shows this call:

```
/*tls_time_server.c continued*/
    SSL_CTX *ctx = SSL_CTX_new(TLS_server_method());
    if (!ctx) {
        fprintf(stderr, "SSL_CTX_new() failed.\n");
        return 1;
    }
```

If you're using an older version of OpenSSL, you may need to replace
TLS_server_method() with TLSv1_2_server_method() in the preceding code.
However, you should probably upgrade to a newer OpenSSL version instead.

Once the SSL context has been created, we can associate our server's certificate with it. The
following code sets the SSL context to use our certificate:

```
/*tls_time_server.c continued*/
    if (!SSL_CTX_use_certificate_file(ctx, "cert.pem" , SSL_FILETYPE_PEM)
        || !SSL_CTX_use_PrivateKey_file(ctx, "key.pem", SSL_FILETYPE_PEM)) {
        fprintf(stderr, "SSL_CTX_use_certificate_file() failed.\n");
        ERR_print_errors_fp(stderr);
        return 1;
    }
```

Make sure that you've generated a proper certificate and key. Refer to the *Self-signed
certificate with OpenSSL* section from earlier in this chapter.

Once the SSL context is configured with the proper certificate, our program creates a
listening TCP socket in the normal way. It begins with a call to getaddrinfo() and
socket(), as shown in the following code:

```
/*tls_time_server.c continued*/

    printf("Configuring local address...\n");
```

```
    struct addrinfo hints;
    memset(&hints, 0, sizeof(hints));
    hints.ai_family = AF_INET;
    hints.ai_socktype = SOCK_STREAM;
    hints.ai_flags = AI_PASSIVE;

    struct addrinfo *bind_address;
    getaddrinfo(0, "8080", &hints, &bind_address);


    printf("Creating socket...\n");
    SOCKET socket_listen;
    socket_listen = socket(bind_address->ai_family,
            bind_address->ai_socktype, bind_address->ai_protocol);
    if (!ISVALIDSOCKET(socket_listen)) {
        fprintf(stderr, "socket() failed. (%d)\n", GETSOCKETERRNO());
        return 1;
    }
```

The socket created by the preceding code is bound to the listening address with `bind()`. The `listen()` function is used to set the socket in a listening state. The following code demonstrates this:

```
/*tls_time_server.c continued*/
    printf("Binding socket to local address...\n");
    if (bind(socket_listen,
                bind_address->ai_addr, bind_address->ai_addrlen)) {
        fprintf(stderr, "bind() failed. (%d)\n", GETSOCKETERRNO());
        return 1;
    }
    freeaddrinfo(bind_address);


    printf("Listening...\n");
    if (listen(socket_listen, 10) < 0) {
        fprintf(stderr, "listen() failed. (%d)\n", GETSOCKETERRNO());
        return 1;
    }
```

If the preceding code isn't familiar, please refer to Chapter 3, *An In-Depth Overview of TCP Connections*.

Note that the preceding code sets the listening port number to 8080. The standard port for HTTPS is 443. It is often more convenient to test with a high port number, since low port numbers require special privileges on some operating systems.

Our server uses a `while` loop to accept multiple connections. Note that this isn't true multiplexing, as only one connection is handled at a time. However, it proves convenient for testing purposes to be able to handle multiple connections serially. Our self-signed certificate causes mainstream browsers to reject our connection on the first try. The connection only succeeds after an exception is added. By having our code loop, it makes adding this exception easier.

Our `while` loop begins by using `accept()` to wait for new connections. This is done by the following code:

```
/*tls_time_server.c continued*/

while (1) {

        printf("Waiting for connection...\n");
        struct sockaddr_storage client_address;
        socklen_t client_len = sizeof(client_address);
        SOCKET socket_client = accept(socket_listen,
                (struct sockaddr*) &client_address, &client_len);
        if (!ISVALIDSOCKET(socket_client)) {
            fprintf(stderr, "accept() failed. (%d)\n", GETSOCKETERRNO());
            return 1;
        }
```

Once the connection is accepted, we use `getnameinfo()` to print out the client's address. This is sometimes useful for debugging purposes. The following code does this:

```
/*tls_time_server.c continued*/

        printf("Client is connected... ");
        char address_buffer[100];
        getnameinfo((struct sockaddr*)&client_address,
                client_len, address_buffer, sizeof(address_buffer), 0, 0,
                NI_NUMERICHOST);
        printf("%s\n", address_buffer);
```

Once the TCP connection is established, an `SSL` object needs to be created. This is done with a call to `SSL_new()`, as shown by the following code:

```
/*tls_time_server.c continued*/

        SSL *ssl = SSL_new(ctx);
        if (!ctx) {
            fprintf(stderr, "SSL_new() failed.\n");
            return 1;
        }
```

The SSL object is associated with the open socket by a call to `SSL_set_fd()`. Then, a TLS/SSL connection can be initialized with a call to `SSL_accept()`. The following code shows this:

```
/*tls_time_server.c continued*/

        SSL_set_fd(ssl, socket_client);
        if (SSL_accept(ssl) <= 0) {
            fprintf(stderr, "SSL_accept() failed.\n");
            ERR_print_errors_fp(stderr);

            SSL_shutdown(ssl);
            CLOSESOCKET(socket_client);
            SSL_free(ssl);

            continue;
        }

        printf ("SSL connection using %s\n", SSL_get_cipher(ssl));
```

In the preceding code, the call to the `SSL_accept()` function can fail for many reasons. For example, if the connected client doesn't trust our certificate, or the client and server can't agree on a cipher suite, then the call to `SSL_accept()` fails. When it fails, we just clean up the allocated resources and use `continue` to repeat our listening loop.

Once the TCP and TLS/SSL connections are fully open, we use `SSL_read()` to receive the client's request. Our program ignores the content of this request. This is because our program only serves the time. It doesn't matter what the client has asked for—our server responds with the time.

The following code uses `SSL_read()` to wait on and read the client's request:

```
/*tls_time_server.c continued*/

        printf("Reading request...\n");
        char request[1024];
        int bytes_received = SSL_read(ssl, request, 1024);
        printf("Received %d bytes.\n", bytes_received);
```

The following code uses `SSL_write()` to transmit the HTTP headers to the client:

```
/*tls_time_server.c continued*/

        printf("Sending response...\n");
        const char *response =
            "HTTP/1.1 200 OK\r\n"
```

```
                "Connection: close\r\n"
                "Content-Type: text/plain\r\n\r\n"
                "Local time is: ";
            int bytes_sent = SSL_write(ssl, response, strlen(response));
            printf("Sent %d of %d bytes.\n", bytes_sent,
    (int)strlen(response));
```

The time() and ctime() functions are then used to format the current time. Once the time is formatted in time_msg, it is also sent to the client using SSL_write(). The following code shows this:

```
/*tls_time_server.c continued*/

            time_t timer;
            time(&timer);
            char *time_msg = ctime(&timer);
            bytes_sent = SSL_write(ssl, time_msg, strlen(time_msg));
            printf("Sent %d of %d bytes.\n", bytes_sent,
    (int)strlen(time_msg));
```

Finally, after the data is transmitted to the client, the connection is closed, and the loop repeats. The following code shows this:

```
/*tls_time_server.c continued*/

            printf("Closing connection...\n");
            SSL_shutdown(ssl);
            CLOSESOCKET(socket_client);
            SSL_free(ssl);
        }
```

If the loop terminates, it would be useful to close the listening socket and clean up the SSL context, as demonstrated by the following code:

```
/*tls_time_server.c continued*/

        printf("Closing listening socket...\n");
        CLOSESOCKET(socket_listen);
        SSL_CTX_free(ctx);
```

Finally, Winsock should be cleaned up if necessary:

```
/*tls_time_server.c continued*/

#if defined(_WIN32)
        WSACleanup();
#endif
```

```
    printf("Finished.\n");

    return 0;
}
```

That concludes `tls_time_server.c`.

You can compile and run the program using the following commands on macOS or Linux:

```
gcc tls_time_server.c -o tls_time_server -lssl -lcrypto
./tls_time_server
```

On Windows, compiling and running the program is done with the following commands:

```
gcc tls_time_server.c -o tls_time_server.exe -lssl -lcrypto -lws2_32
tls_time_server
```

If you have linker errors, please be sure that the OpenSSL library is installed correctly. You may find it helpful to attempt to compile `openssl_version.c` from Chapter 9, *Loading Secure Web Pages with HTTPS and OpenSSL*.

The following screenshot shows what running `tls_time_server` might look like:

You can connect to the time server by navigating your web browser to
`https://127.0.0.1:8080`. Upon the first connection, your browser will reject the self-
signed certificate. The following screenshot shows what this rejection looks like in Firefox:

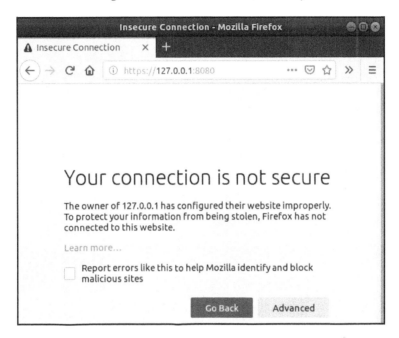

To access the time server, you need to add an exception in the browser. The method for this
is different in each browser, but generally, there is an **Advanced** button that leads to an
option to either add a certificate exception or otherwise proceed with the insecure
connection.

Once the browser connection is established, you will be able to see the current time as given
by our `tls_time_server` program:

The `tls_time_server` program proved useful to show how a TLS/SSL server can be set up without getting bogged down in the details of actualizing a complete HTTPS server. However, this chapter's code repository also includes a more substantial HTTPS server.

A full HTTPS server

Included in this chapter's code repository is `https_server.c`. This program is a modification of `web_server.c` from Chapter 7, *Building a Simple Web Server*. It can be used to serve a simple static website over HTTPS.

In the `https_server.c` program, the basic TLS/SSL connection is set up and established the same way as shown in `tls_time_server.c`. Once the secure connection is established, the connection is simply treated as HTTP.

`https_server` is compiled using the same technique as for `tls_time_server`. The following screenshot shows how to compile and run `https_server`:

```
honp@ubby18: ~/Desktop
$ gcc https_server.c -o https_server -lssl -lcrypto
$ ./https_server
Configuring local address...
Creating socket...
Binding socket to local address...
Listening...
New connection from 127.0.0.1.
SSL connection using ECDHE-RSA-AES128-GCM-SHA256
serve_resource 127.0.0.1 /
New connection from 127.0.0.1.
SSL connection using ECDHE-RSA-AES128-GCM-SHA256
serve_resource 127.0.0.1 /smile.png
```

Once `https_server` is running, you can connect to it by navigating your web browser to `https://127.0.0.1:8080`. You will likely need to add a security exception when connecting for the first time. The code is set up to serve the static pages from the `chap07` directory.

The following screenshot was taken of a web browser connected to `https_server`:

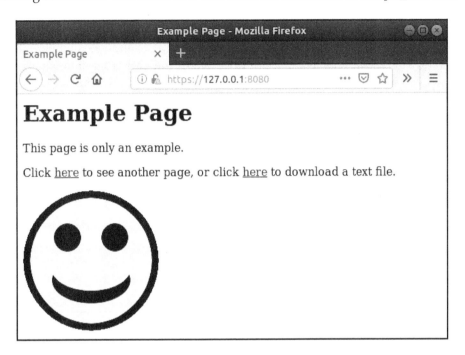

This chapter's example programs illustrate the basics of HTTPS servers. However, implementing a genuinely robust HTTPS server does involve additional challenges. Let's consider some of these now.

HTTPS server challenges

This chapter should serve only as an introduction to TLS/SSL server programming. There is much more to learn about secure network programming. Before deploying a secure HTTPS server with OpenSSL, it is essential to review all the OpenSSL documentation carefully. Many OpenSSL functions have edge cases that were ignored in the illustrative code for this chapter.

Multiplexing can also be complicated with OpenSSL. In typical TCP servers, we have been using the `select()` function to indicate when data is available to be read.

The `select()` function works directly on the TCP socket. Using `select()` on a server secured with TLS/SSL can be tricky. This is because `select()` indicates when data is available at the TCP level. This usually, but not always, indicates that data is available to be read with `SSL_read()`. It is important that you carefully consult the OpenSSL documentation for `SSL_read()` if you are going to use it with `select()`. The example program in this chapter ignores these possibilities for reasons of simplicity.

There are also alternatives to OpenSSL. Let's consider some of them now.

OpenSSL alternatives

Although **OpenSSL** is one of the oldest and most widely deployed libraries implementing TLS, many alternative libraries have sprung up in recent years. Some of these alternatives aim to offer better features, performance, or quality control compared to OpenSSL.

The following table contains a number of alternative open source TLS libraries:

TLS Library	Website
cryptlib	https://www.cryptlib.com/
GnuTLS	https://www.gnutls.org/
LibreSSL	https://www.libressl.org/
mbed TLS	https://tls.mbed.org/
Network Security Services	https://developer.mozilla.org/en-US/docs/Mozilla/Projects/NSS
s2n	https://github.com/awslabs/s2n
wolfSSL	https://www.wolfssl.com/

There are also alternatives to doing TLS termination directly in your application, and this can simplify program design. Let's consider this next.

Alternatives to TLS

Getting everything right when implementing an HTTPS server can prove difficult, and missing even a minor detail can compromise security entirely.

As an alternative to using TLS directly in our server itself, it is sometimes a better idea to use a **reverse proxy server**. A reverse proxy server can be configured to accept secure connections from clients and then forward these connections as plain HTTP to your program.

Nginx and **Apache** are two popular open source servers that can work well as HTTPS reverse proxies.

This setup is illustrated by the following diagram:

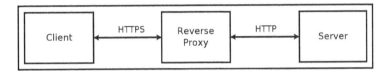

A reverse proxy server configured in this way is also called a **TLS termination proxy**.

An even better alternative may be to create your program using the **CGI** or **FastCGI** standard. In this case, your program communicates directly with a standard web server. The web server handles all of the HTTPS and HTTP details. This can greatly simplify program design, and, in some cases, reduce maintenance costs too.

If you do use an off-the-shelf HTTPS server, it is still important to use caution. It can be easy to inadvertently compromise security through misconfiguration.

Summary

In this chapter, we considered the HTTPS protocol from the server's perspective. We covered how certificates work, and we showed the method for generating a self-signed certificate with OpenSSL.

Once we had a certificate, we learned how to use the OpenSSL library to listen for TLS/SSL connections. We used this knowledge to implement a simple server that displays the current time over HTTPS.

We also discussed some of the pitfalls and complexity of implementing HTTPS servers. Many applications may benefit from side-stepping the implementation of HTTPS and relying on a reverse proxy instead.

In the next chapter, Chapter 11, *Establishing SSH Connections with libssh*, we will look at another secure protocol, **Secure Shell** (**SSH**).

Questions

Try these questions to test the knowledge you have acquired from this chapter:

1. How does a client decide whether it should trust a server's certificate?
2. What is the main issue with self-signed certificates?
3. What can cause `SSL_accept()` to fail?
4. Can `select()` be used to multiplex connections for HTTPS servers?

The answers to these questions can be found in `Appendix A`, *Answers to Questions*.

Further reading

For more information about HTTPS and OpenSSL, please refer to the following:

- OpenSSL documentation (`https://www.openssl.org/docs/`)
- **RFC 5246**: *The **Transport Layer Security** (TLS) Protocol Version 1.2* (`https://tools.ietf.org/html/rfc5246`)
- *Let's Encrypt* (`https://letsencrypt.org/`)

11
Establishing SSH Connections with libssh

This chapter is all about programming with the **Secure Shell** (**SSH**) protocol. SSH is a secure network protocol used to authenticate with remote servers, grant command-line access, and securely transfer files.

SSH is widely used for the configuration and management of remote servers. Oftentimes, web servers aren't connected to monitors or keyboards. For many of these servers, SSH provides the only method of command-line access and administration.

The following topics are covered in this chapter:

- SSH protocol overview
- The `libssh` library
- Establishing a connection
- SSH authentication methods
- Executing a remote command
- File transfers

Technical requirements

The example programs from this chapter can be compiled using any modern C compiler. We recommend **MinGW** for Windows and **GCC** for Linux and macOS. You also need to have the `libssh` library installed. See `Appendix B`, *Setting Up Your C Compiler on Windows*, `Appendix C`, *Setting Up Your C Compiler on Linux*, and `Appendix D`, *Setting Up Your C Compiler on macOS*, for compiler setup and `libssh` installation.

The code for this book can be found at `https://github.com/codeplea/Hands-On-Network-Programming-with-C`.

From the command-line, you can download the code for this chapter by using the following command:

```
git clone https://github.com/codeplea/Hands-On-Network-Programming-with-C
cd Hands-On-Network-Programming-with-C/chap11
```

Each example program in this chapter runs on Windows, Linux, and macOS.

Each example needs to be linked against the libssh library. This is accomplished by passing the -lssh option to gcc.

We provide the exact commands needed to compile each example as it is introduced.

For brevity, we use a standard header file with each example program in this chapter. This header includes the other needed headers in one place. Its contents are as follows:

```
/*chap11.h*/

#include <stdio.h>
#include <stdlib.h>
#include <string.h>

#include <libssh/libssh.h>
```

The SSH protocol

Most servers providing a service (such as websites and emails) over the modern internet aren't attached to keyboards or monitors. Even when servers do have local input/output hardware, remote access is often much more convenient.

Various protocols have been used to provide remote command-line access to servers. One of the first such protocols was **Telnet**. With Telnet, a client remotely connects to a server using plaintext over TCP port 23. The server provides more-or-less direct access to the operating system command-line through this **Transmission Control Protocol** (TCP) connection. The client sends plaintext commands to the server, and the server executes these commands. The command-line output is sent back from the server to the client.

Telnet has a major security shortcoming: it does not encrypt any data sent over the network. Even user passwords are sent as plaintext when using Telnet. This means that any network eavesdropper could obtain user credentials!

The SSH protocol has now largely replaced Telnet. The SSH protocol works over TCP using port 22. SSH uses strong encryption to protect against eavesdropping.

SSH allows clients to verify servers' identities using **public-key authentication**. Without public-key authentication of the server, an impostor could masquerade as a legitimate server and attempt to trick a client into connecting. Once connected, the client would send its credentials to the impostor server.

SSH also provides many methods for client authentication with severs. These include sending a password or using public-key authentication. We look at these methods in detail later.

SSH is a complicated protocol. So, instead of attempting to implement it ourselves, we use an existing library to provide the needed functionality.

libssh

libssh is a widely used open source C library implementing the SSH protocol. It allows us to remotely execute commands and transfer files using the SSH protocol.

libssh is structured in a way that abstracts network connections. We won't need to bother with the low-level networking APIs we've been using so far. The libssh library handles hostname resolution and creation of the needed TCP sockets for us.

Testing out libssh

Before continuing with this chapter, it is essential that you have the libssh library installed and available. Please refer to Appendix B, *Setting Up Your C Compiler on Windows*, Appendix C, *Setting Up Your C Compiler on Linux*, and Appendix D, *Setting Up Your C Compiler on macOS*, for libssh installation.

Our first program using libssh is designed to ensure that it's installed correctly. This program merely prints the libssh library version. The program is as follows:

```
/*ssh_version.c*/

#include "chap11.h"

int main()
{
    printf("libssh version: %s\n", ssh_version(0));
    return 0;
}
```

You can compile and run `ssh_version.c` with the following commands on Windows using MinGW:

```
gcc ssh_version.c -o ssh_version.exe -lssh
ssh_version
```

On Linux and macOS, the commands to compile and run `ssh_version.c` are as follows:

```
gcc ssh_version.c -o ssh_version -lssh
./ssh_version
```

The following screenshot shows `ssh_version.c` being successfully compiled and run on Linux:

If you receive an error message about `libssh.h` not being found, you should check that you have the `libssh` library headers in your compiler's `include` directory search path. If you see an error message about an undefined reference to `ssh_version`, then please check that you didn't forget to pass the `-lssh` option to your compiler.

The next step to understanding `libssh` is to establish an actual SSH connection.

Establishing a connection

Now that we've ensured that `libssh` is correctly installed, it's time to attempt an actual SSH connection.

You'll need to have access to an SSH server before continuing. OpenSSH is a popular server that is available for Linux, macOS, and Windows 10. It works well for testing but be sure to understand the security implementations before installing it on your device. Refer to your operating system's documentation for more information.

If you would rather test with a remote system, Linux **Virtual Private Servers** (**VPS**) running OpenSSH are available from many providers. They typically cost only a few dollars a month.

Let's continue by implementing a program that uses `libssh` to open an SSH connection.

We structure the rest of the programs in this chapter to take the SSH server's hostname and port number as command-line arguments. Our program starts with the following code, which checks these arguments:

```
/*ssh_connect.c*/

#include "chap11.h"

int main(int argc, char *argv[])
{
    const char *hostname = 0;
    int port = 22;
    if (argc < 2) {
        fprintf(stderr, "Usage: ssh_connect hostname port\n");
        return 1;
    }
    hostname = argv[1];
    if (argc > 2) port = atol(argv[2]);
```

In the preceding code, `argc` is checked to see whether at least the hostname was passed in as a command-line argument. If it wasn't, a usage message is displayed instead. Otherwise, the server's hostname is stored in the `hostname` variable. If a port number was passed in, it is stored in the `port` variable. Otherwise, the default port `22` is stored instead.

SSH often provides complete and total access to a server. For this reason, some internet criminals randomly scan IP addresses for SSH connections. When they successfully establish a connection, they attempt to guess login credentials, and if successful, they take control of the server. These attacks aren't successful against properly secured servers, but they are a common nuisance, nonetheless. Using SSH on a port other than the default (`22`) often avoids these automated attacks. This is one reason why we want to ensure our programs work well with non-default port numbers.

Once our program has obtained the hostname and connection port number, we continue by creating an SSH session object. This is done with a call to ssh_new() as shown in the following code:

```
/*ssh_connect.c continued*/

    ssh_session ssh = ssh_new();
    if (!ssh) {
        fprintf(stderr, "ssh_new() failed.\n");
        return 1;
    }
```

The preceding code creates a new SSH session object and stores it in the ssh variable.

Once the SSH session is created, we need to specify some options before completing the connection. The ssh_options_set() function is used to set options. The following code shows setting the remote hostname and port:

```
/*ssh_connect.c continued*/

    ssh_options_set(ssh, SSH_OPTIONS_HOST, hostname);
    ssh_options_set(ssh, SSH_OPTIONS_PORT, &port);
```

libssh includes useful debugging tools. By setting the SSH_OPTIONS_LOG_VERBOSITY option, we tell libssh to print almost everything it does. The following code causes libssh to log a lot of information about which actions it takes:

```
/*ssh_connect.c continued*/

    int verbosity = SSH_LOG_PROTOCOL;
    ssh_options_set(ssh, SSH_OPTIONS_LOG_VERBOSITY, &verbosity);
```

This logging is useful to see, but it can also be distracting. I recommend you try it once and then disable it unless you run into problems. The rest of this chapter's examples won't use it.

We can now use ssh_connect() to initiate the SSH connection. The following code shows this:

```
/*ssh_connect.c continued*/

    int ret = ssh_connect(ssh);
    if (ret != SSH_OK) {
        fprintf(stderr, "ssh_connect() failed.\n%s\n", ssh_get_error(ssh));
        return -1;
    }
```

Note that `ssh_connect()` returns `SSH_OK` on success. On failure, we use the `ssh_get_error()` function to detail what went wrong.

Next, our code prints out that the connection was successful:

```
/*ssh_connect.c continued*/

    printf("Connected to %s on port %d.\n", hostname, port);
```

The SSH protocol allows servers to send a message to clients upon connecting. This message is called the **banner**. It is typically used to identify the server or provide short access rules. We can print the banner using the following code:

```
/*ssh_connect.c continued*/

    printf("Banner:\n%s\n", ssh_get_serverbanner(ssh));
```

This is as far as our `ssh_connect.c` example goes. Our program simply disconnects and frees the SSH session before terminating. The following code concludes `ssh_connect.c`:

```
/*ssh_connect.c continued*/

    ssh_disconnect(ssh);
    ssh_free(ssh);

    return 0;
}
```

You can compile `ssh_connect.c` with the following command on Windows using MinGW:

gcc ssh_connect.c -o ssh_connect.exe -lssh

On Linux and macOS, the command to compile `ssh_connect.c` is as follows:

gcc ssh_connect.c -o ssh_connect -lssh

The following screenshot shows `ssh_connect.c` being successfully compiled and run on Linux:

```
honp@ubby18: ~/Desktop                                                      ─ ▢ ⊗
$ gcc ssh_connect.c -o ssh_connect -lssh
$ ./ssh_connect localhost
[2019/02/15 19:44:44.037146, 2] ssh_connect:  libssh 0.8.6 (c) 2003-2018 Aris Adamantiadis, Andre
as Schneider and libssh contributors. Distributed under the LGPL, please refer to COPYING file fo
r information about your rights, using threading threads_pthread
[2019/02/15 19:44:44.037397, 2] ssh_socket_connect:  Nonblocking connection socket: 3
[2019/02/15 19:44:44.037414, 2] ssh_connect:  Socket connecting, now waiting for the callbacks to
 work
[2019/02/15 19:44:44.037431, 1] socket_callback_connected:  Socket connection callback: 1 (0)
[2019/02/15 19:44:44.275512, 1] ssh_client_connection_callback:  SSH server banner: SSH-2.0-OpenS
SH_7.6p1 Ubuntu-4ubuntu0.2
[2019/02/15 19:44:44.275561, 1] ssh_analyze_banner:  Analyzing banner: SSH-2.0-OpenSSH_7.6p1 Ubun
tu-4ubuntu0.2
[2019/02/15 19:44:44.275594, 1] ssh_analyze_banner:  We are talking to an OpenSSH client version:
 7.6 (70600)
[2019/02/15 19:44:44.342304, 1] ssh_known_hosts_read_entries:  Failed to open the known_hosts fil
e '/etc/ssh/ssh_known_hosts': No such file or directory
[2019/02/15 19:44:44.352247, 2] ssh_kex_select_methods:  Negotiated curve25519-sha256,ecdsa-sha2-
nistp256,aes256-ctr,aes256-ctr,hmac-sha2-256,hmac-sha2-256,none,none,,
[2019/02/15 19:44:44.387545, 2] ssh_packet_dh_reply:  Received SSH_KEXDH_REPLY
[2019/02/15 19:44:44.402617, 2] ssh_client_curve25519_reply:  SSH_MSG_NEWKEYS sent
[2019/02/15 19:44:44.402662, 2] ssh_packet_newkeys:  Received SSH_MSG_NEWKEYS
[2019/02/15 19:44:44.440037, 2] ssh_packet_newkeys:  Signature verified and valid
Connected to localhost on port 22.
Banner:
SSH-2.0-OpenSSH_7.6p1 Ubuntu-4ubuntu0.2
$ ▊
```

In the preceding screenshot, you can see that `ssh_connect` was able to connect to the OpenSSH server running locally.

Now that we've established a connection, let's continue by authenticating with the server.

SSH authentication

SSH provides authentication methods for both the server (host) and the client (user). It should be obvious why the server must authenticate the client. The server wants to only provide access to authorized users. Otherwise, anyone could take over the server.

However, the client also needs to authenticate the server. If the client fails to authenticate the server properly, then the client could be tricked into sending its password to an impostor!

In SSH, servers are authenticated using public key encryption. Conceptually, this is very similar to how HTTPS provides server authentication. However, SSH doesn't typically rely on certificate authorities. Instead, when using SSH, most clients simply keep a list of the public keys (or hashes of the public keys) that they trust. How the clients obtain this list in the first place varies. Generally, if a client connects to a server under trusted circumstances, then it can trust that public key in the future too.

libssh implements features to remember trusted servers' public keys. In this way, once a server has been connected to and trusted once, libssh remembers that it's trusted in the future.

Some SSH deployments also use other methods to validate SSH hosts' public keys. For example, a **Secure Shell fingerprint** (**SSHFP**) record is a type of DNS record used to validate SSH public keys. Its use requires secure DNS access.

Regardless of how you decide to trust (or not trust) a server's public key, you'll need to obtain the server's public key in the first place. Let's look at how libssh provides access to the server authentication functionality.

Server authentication

Once the SSH session is established, we can get the server's public key using the ssh_get_server_publickey() function. The following code illustrates this function call:

```
/*ssh_auth.c excerpt*/

    ssh_key key;
    if (ssh_get_server_publickey(ssh, &key) != SSH_OK) {
        fprintf(stderr, "ssh_get_server_publickey() failed.\n%s\n",
                ssh_get_error(ssh));
        return -1;
    }
```

It is often useful to obtain and display a hash of the server's SSH public key. Users can look at hashes and compare these to known keys. The libssh library provides the ssh_get_publickey_hash() function for this purpose.

The following code prints out an SHA1 hash of the public key obtained earlier:

```
/*ssh_auth.c excerpt*/

    unsigned char *hash;
    size_t hash_len;
    if (ssh_get_publickey_hash(key, SSH_PUBLICKEY_HASH_SHA1,
```

```
                    &hash, &hash_len) != SSH_OK) {
        fprintf(stderr, "ssh_get_publickey_hash() failed.\n%s\n",
                ssh_get_error(ssh));
        return -1;
    }

    printf("Host public key hash:\n");
    ssh_print_hash(SSH_PUBLICKEY_HASH_SHA1, hash, hash_len);
```

libssh prints SHA1 hashes using Base64. It also prepends the hash type first. For example, the preceding code might print the following:

```
Host public key hash:
SHA1:E348CMNeCGGec/bQqEX7aocDTfI
```

When you've finished with the public key and hash, free their resources with the following code:

```
/*ssh_auth.c excerpt*/

    ssh_clean_pubkey_hash(&hash);
    ssh_key_free(key);
```

libssh provides the ssh_session_is_known_server() function to determine whether a server's public key is known. The following code shows an example of using this code:

```
/*ssh_auth.c excerpt*/

    enum ssh_known_hosts_e known = ssh_session_is_known_server(ssh);
    switch (known) {
        case SSH_KNOWN_HOSTS_OK: printf("Host Known.\n"); break;

        case SSH_KNOWN_HOSTS_CHANGED: printf("Host Changed.\n"); break;
        case SSH_KNOWN_HOSTS_OTHER: printf("Host Other.\n"); break;
        case SSH_KNOWN_HOSTS_UNKNOWN: printf("Host Unknown.\n"); break;
        case SSH_KNOWN_HOSTS_NOT_FOUND: printf("No host file.\n"); break;

        case SSH_KNOWN_HOSTS_ERROR:
            printf("Host error. %s\n", ssh_get_error(ssh)); return 1;

        default: printf("Error. Known: %d\n", known); return 1;
    }
```

If the server's public key is known (previously trusted), then
`ssh_session_is_known_server()` returns `SSH_KNOWN_HOSTS_OK`. Otherwise,
`ssh_session_is_known_server()` can return other values with various meanings.

`SSH_KNOWN_HOSTS_UNKNOWN` indicates that the server is unknown. In this case, the user
should verify the server's hash.

`SSH_KNOWN_HOSTS_NOT_FOUND` means that `libssh` didn't find a hosts file, and one is
created automatically. This should generally be treated in the same way as
`SSH_KNOWN_HOSTS_UNKNOWN`.

`SSH_KNOWN_HOSTS_CHANGED` indicates that the server is returning a different key than was
previously known, while `SSH_KNOWN_HOSTS_OTHER` indicates that the server is returning a
different type of key than was previously used. Either of these could indicate a potential
attack! In a real-world application, you should be more explicit about notifying the user of
these risks.

If the user has verified that a host is to be trusted, use
`ssh_session_update_known_hosts()` to allow `libssh` to save the servers public key
hash. This allows `ssh_session_is_known_server()` to return `SSH_KNOWN_HOSTS_OK`
for the next connection.

The following code illustrates prompting the user to trust a connection and using
`ssh_session_update_known_hosts()`:

```
/*ssh_auth.c excerpt*/

    if (known == SSH_KNOWN_HOSTS_CHANGED ||
            known == SSH_KNOWN_HOSTS_OTHER ||
            known == SSH_KNOWN_HOSTS_UNKNOWN ||
            known == SSH_KNOWN_HOSTS_NOT_FOUND) {
        printf("Do you want to accept and remember this host? Y/N\n");
        char answer[10];
        fgets(answer, sizeof(answer), stdin);
        if (answer[0] != 'Y' && answer[0] != 'y') {
            return 0;
        }

        ssh_session_update_known_hosts(ssh);
    }
```

Please see `ssh_auth.c` in this chapter's code repository for a working example. Consult the
`libssh` documentation for more information.

After the client has authenticated the server, the server needs to authenticate the client.

Client authentication

SSH offers several methods of client authentication. These methods include the following:

- **No authentication**: This allows any user to connect
- **Password authentication**: This requires the user to provide a username and password
- **Public key**: This uses public key encryption methods to authenticate
- **Keyboard-interactive**: This authenticates by having the user answer several prompts
- **Generic Security Service Application Program Interface (GSS-API)**: This allows authentication through a variety of other services

Password authentication is the most common method, but it does have some drawbacks. If an impostor server tricks a user into sending their password, then that user's password is effectively compromised. Public key user authentication doesn't suffer from this attack to the same degree. With public key authentication, the server issues a unique challenge for each authentication attempt. This prevents a malicious impostor server from replaying a previous authentication to the legitimate server.

Once public key authentication is set up, libssh makes using it very simple. In many cases, it's as easy as calling the ssh_userauth_publickey_auto() function. However, setting up public key authentication in the first place can be a tedious process.

Although public key authentication is more secure, password authentication is still in common use. Password authentication is also more straightforward and easier to test. For these reasons, we continue the examples in this chapter by using password authentication.

Regardless of the user authentication method, the SSH server must know what user you are trying to authenticate as. The libssh library lets us provide this information using the ssh_set_options() function that we saw earlier. It should be called before using ssh_connect(). To set the user, call ssh_options_set() with SSH_OPTIONS_USER as shown in the following code:

```
ssh_options_set(ssh, SSH_OPTIONS_USER, "alice");
```

After the SSH session has been established, a password can be provided with the
`ssh_userauth_password()` function. The following code prompts for a password and
sends it to the connected SSH server:

```
/*ssh_auth.c excerpt*/

    printf("Password: ");
    char password[128];
    fgets(password, sizeof(password), stdin);
    password[strlen(password)-1] = 0;

    if (ssh_userauth_password(ssh, 0, password) != SSH_AUTH_SUCCESS) {
        fprintf(stderr, "ssh_userauth_password() failed.\n%s\n",
                ssh_get_error(ssh));
        return 0;
    } else {
        printf("Authentication successful!\n");
    }
```

The preceding code uses the `fgets()` function to obtain the password from the user.
The `fgets()` function always includes the newline character with the input, which we
don't want. The `password[strlen(password)-1] = 0` code effectively shortens the
password by one character, thus removing the newline character.

Note that using `fgets()` causes the entered password to display on the screen. This isn't
secure, and it would be an improvement to hide the password while it's being entered.
Unfortunately, there isn't a cross-platform way to do this. If you're using Linux, consider
using the `getpass()` function in place of `fgets()`.

See `ssh_auth.c` in this chapter's code repository for a working example of authenticating
with a server using user password authentication.

You can compile and run `ssh_auth.c` with the following commands on Windows using
MinGW:

```
gcc ssh_auth.c -o ssh_auth.exe -lssh
ssh_auth example.com 22 alice
```

On Linux and macOS, the commands to compile and run `ssh_auth.c` are as follows:

```
gcc ssh_auth.c -o ssh_auth -lssh
./ssh_auth example.com 22 alice
```

The following screenshot shows compiling `ssh_auth` and using it to connect to a locally running SSH server on Linux:

In the preceding screenshot, `ssh_auth` was used to successfully authenticate with the locally running SSH server. The `ssh_auth` program used password authentication with the username `alice` and the password `password123`. Needless to say, you need to change the username and password as appropriate for your SSH server. Authentication will be successful only if you use the username and password for an actual user account on the server you connect to.

After authenticating, we're ready to run a command over SSH.

Executing a remote command

The SSH protocol works using channels. After we've established an SSH connection, a channel must be opened to do any real work. The advantage is that many channels can be opened over one connection. This potentially allows an application to do multiple things (seemingly) simultaneously.

After the SSH session is open and the user is authenticated, a channel can be opened. A new channel is opened by calling the `ssh_channel_new()` function. The following code illustrates this:

```
/*ssh_command.c excerpt*/

    ssh_channel channel = ssh_channel_new(ssh);
    if (!channel) {
        fprintf(stderr, "ssh_channel_new() failed.\n");
        return 0;
    }
```

The SSH protocol implements many types of channels. The **session** channel type is used for executing remote commands and transferring files. With `libssh`, we can request a session channel by using the `ssh_channel_open_session()` function. The following code shows calling `ssh_channel_open_session()`:

```
/*ssh_command.c excerpt*/

    if (ssh_channel_open_session(channel) != SSH_OK) {
        fprintf(stderr, "ssh_channel_open_session() failed.\n");
        return 0;
    }
```

Once the session channel is open, we can issue a command to run with the `ssh_channel_request_exec()` function. The following code uses `fgets()` to prompt the user for a command and `ssh_channel_request_exec()` to send the command to the remote host:

```
/*ssh_command.c excerpt*/

    printf("Remote command to execute: ");
    char command[128];
    fgets(command, sizeof(command), stdin);
    command[strlen(command)-1] = 0;

    if (ssh_channel_request_exec(channel, command) != SSH_OK) {
        fprintf(stderr, "ssh_channel_open_session() failed.\n");
        return 1;
    }
```

Once the command has been sent, our program uses `ssh_channel_read()` to receive the command output. The following code loops until the entire output is read:

```
/*ssh_command.c excerpt*/

    char output[1024];
```

```
int bytes_received;
while ((bytes_received =
            ssh_channel_read(channel, output, sizeof(output), 0))) {
    if (bytes_received < 0) {
        fprintf(stderr, "ssh_channel_read() failed.\n");
        return 1;
    }
    printf("%.*s", bytes_received, output);
}
```

The preceding code first allocates a buffer, `output`, to hold the received data from the command's output. The `ssh_channel_read()` function returns the number of bytes read, but it returns `0` when the read is complete or a negative number for an error. Our code loops while `ssh_channel_read()` returns data.

After the entire output from the command has been received, the client should send an **end-of-file** (**EOF**) over the channel, close the channel, and free the channel resources. The following code shows this:

```
/*ssh_command.c excerpt*/

ssh_channel_send_eof(channel);
ssh_channel_close(channel);
ssh_channel_free(channel);
```

If your program is also done with the SSH session, be sure to call `ssh_disconnect()` and `ssh_free()` as well.

The `ssh_command.c` program included in this chapter's code repository is a simple utility that connects to a remote SSH host and executes a single command.

You can compile `ssh_command.c` with the following command on Windows using MinGW:

```
gcc ssh_command.c -o ssh_command.exe -lssh
```

On Linux and macOS, the command to compile `ssh_command.c` is as follows:

```
gcc ssh_command.c -o ssh_command -lssh
```

The following screenshot shows `ssh_command.c` being compiled and run on Linux:

```
                    honp@ubby18: ~/Desktop                    ⊖ ⊡ ⊗
$ gcc ssh_command.c -o ssh_command -lssh
$ ./ssh_command localhost 22 alice
Connected to localhost on port 22.
Banner:
SSH-2.0-OpenSSH_7.6p1 Ubuntu-4ubuntu0.2
Host public key hash:
SHA1:UWXxDb3ArslAr7i2YBW07rLugtE
Checking ssh_session_is_known_server()
Host Known.
Password: password123
Authentication successful!
Remote command to execute: ls -l
total 24
-rw-rw-r-- 1 alice alice  312 Feb 15 07:52 credits.txt
-rw-r--r-- 1 alice alice 8980 Apr 16  2018 examples.desktop
-rw-rw-r-- 1 alice alice 2124 Feb 15 07:51 schedule.txt
-rw-rw-r-- 1 alice alice   35 Feb 15 07:51 test.txt
ssh_channel_read() failed.
$
```

The preceding screenshot shows connecting to the local OpenSSH server and executing the `ls -l` command. The `ssh_command` code faithfully prints the output of that command (which is a file listing for the user's home directory).

The `libssh` library function `ssh_channel_request_exec()` is useful to execute a single command. However, SSH also supports methods for opening a fully interactive remote shell. Generally, a session channel is opened as shown previously. Then the `libssh` library function `ssh_channel_request_pty()` is called to initialize a remote shell. The `libssh` library provides many functions to send and receive data this way. Please refer to the `libssh` documentation for more information.

Now that you're able to execute a remote command and receive its output, it may also be useful to transfer files. Let's consider that next.

Downloading a file

The **Secure Copy Protocol** (**SCP**) provides a method to transfer files. It supports both uploading and downloading files.

`libssh` makes using SCP easy. This chapter's code repository contains an example, `ssh_download.c`, which shows the basic method for downloading a file over SCP with `libssh`.

After the SSH session is started and the user is authenticated, `ssh_download.c` prompts the user for the remote filename using the following code:

```
/*ssh_download.c excerpt*/

    printf("Remote file to download: ");
    char filename[128];
    fgets(filename, sizeof(filename), stdin);
    filename[strlen(filename)-1] = 0;
```

A new SCP session is initialized by calling the `libssh` library function `ssh_scp_new()`, as shown in the following code:

```
/*ssh_download.c excerpt*/

    ssh_scp scp = ssh_scp_new(ssh, SSH_SCP_READ, filename);
    if (!scp) {
        fprintf(stderr, "ssh_scp_new() failed.\n%s\n",
                ssh_get_error(ssh));
        return 1;
    }
```

In the preceding code, `SSH_SCP_READ` is passed to `ssh_scp_new()`. This specifies that we are going to use the new SCP session for downloading files. The `SSH_SCP_WRITE` option would be used to upload files. The `libssh` library also provides the `SSH_SCP_RECURSIVE` option to assist with uploading or downloading entire directory trees.

After the SCP session is created successfully, `ssh_scp_init()` must be called to initialize the SCP channel. The following code shows this:

```
/*ssh_download.c excerpt*/

    if (ssh_scp_init(scp) != SSH_OK) {
        fprintf(stderr, "ssh_scp_init() failed.\n%s\n",
                ssh_get_error(ssh));
        return 1;
    }
```

`ssh_scp_pull_request()` must be called to begin the file download. This function returns `SSH_SCP_REQUEST_NEWFILE` to indicate that the remote host is going to begin sending a new file. The following code shows this:

```
/*ssh_download.c excerpt*/

    if (ssh_scp_pull_request(scp) != SSH_SCP_REQUEST_NEWFILE) {
```

```
        fprintf(stderr, "ssh_scp_pull_request() failed.\n%s\n",
                ssh_get_error(ssh));
        return 1;
    }
```

libssh provides some methods we can use to retrieve the remote filename, file size, and permissions. The following code retrieves these values and prints them to the console:

```
/*ssh_download.c excerpt*/

    int fsize = ssh_scp_request_get_size(scp);
    char *fname = strdup(ssh_scp_request_get_filename(scp));
    int fpermission = ssh_scp_request_get_permissions(scp);

    printf("Downloading file %s (%d bytes, permissions 0%o\n",
           fname, fsize, fpermission);
    free(fname);
```

Once the file size is known, we can allocate space to store it in memory. This is done using malloc() as shown in the following code:

```
/*ssh_download.c excerpt*/

    char *buffer = malloc(fsize);
    if (!buffer) {
        fprintf(stderr, "malloc() failed.\n");
        return 1;
    }
```

Our program then accepts the new file request with ssh_scp_accept_request() and downloads the file with ssh_scp_read(). The following code shows this:

```
/*ssh_download.c excerpt*/

    ssh_scp_accept_request(scp);
    if (ssh_scp_read(scp, buffer, fsize) == SSH_ERROR) {
        fprintf(stderr, "ssh_scp_read() failed.\n%s\n",
                ssh_get_error(ssh));
        return 1;
    }
```

The downloaded file can be printed to the screen with a simple call to `printf()`. When we're finished with the file data, it's important to free the allocated space too. The following code prints out the file's contents and frees the allocated memory:

```
/*ssh_download.c excerpt*/

    printf("Received %s:\n", filename);
    printf("%.*s\n", fsize, buffer);
    free(buffer);
```

An additional call to `ssh_scp_pull_request()` should return `SSH_SCP_REQUEST_EOF`. This indicates that we received the entire file from the remote host. The following code checks for the end-of-file request from the remote host:

```
/*ssh_download.c excerpt*/

    if (ssh_scp_pull_request(scp) != SSH_SCP_REQUEST_EOF) {
        fprintf(stderr, "ssh_scp_pull_request() unexpected.\n%s\n",
                ssh_get_error(ssh));
        return 1;
    }
```

The preceding code is simplified a bit. The remote host could also return other values which aren't necessarily errors. For example, if `ssh_scp_pull_request()` returns `SSH_SCP_REQUEST_WARNING`, then the remote host has sent a warning. This warning can be read by calling `ssh_scp_request_get_warning()`, but, in any case, `ssh_scp_pull_request()` should be called again.

After the file is received, `ssh_scp_close()` and `ssh_scp_free()` should be used to free resources as shown in the following code excerpt:

```
/*ssh_download.c excerpt*/

    ssh_scp_close(scp);
    ssh_scp_free(scp);
```

After your program is done with the SSH session, don't forget to call `ssh_disconnect()` and `ssh_free()` as well.

The entire file-downloading example is included with this chapter's code as `ssh_download.c`.

You can compile `ssh_download.c` with the following command on Windows using MinGW:

```
gcc ssh_download.c -o ssh_download.exe -lssh
```

On Linux and macOS, the command to compile `ssh_download.c` is as follows:

```
gcc ssh_download.c -o ssh_download -lssh
```

The following screenshot shows `ssh_download.c` being successfully compiled and used on Linux to download a file:

As you can see from the preceding screenshot, downloading a file using SSH and SCP is pretty straightforward. This can be a useful way to transfer data between computers securely.

Summary

This chapter provided a cursory overview of the SSH protocol and how to use it with libssh. We learned a lot about authentication with the SSH protocol, and how the server and client must both authenticate for security. Once the connection was established, we implemented a simple program to execute a command on a remote host. We also saw how libssh makes downloading a file using SCP very easy.

SSH provides a secure communication channel, which effectively denies eavesdroppers the meaning of intercepted communication.

In the next chapter, Chapter 12, *Network Monitoring and Security*, we continue with the security theme by looking at tools that can effectively eavesdrop on non-secure communication channels.

Questions

Try these questions to test your knowledge from this chapter:

1. What is a significant downside of using Telnet?
2. Which port does SSH typically run on?
3. Why is it essential that the client authenticates the SSH server?
4. How is the server typically authenticated?
5. How is the SSH client typically authenticated?

The answers to this questions can be found in Appendix A, *Answers to Questions*.

Further reading

For more information about Telnet, SSH, and `libssh`, please refer to the following:

- The SSH library `https://www.libssh.org`
- **RFC 15**: *Network Subsystem for Time Sharing Hosts* (Telnet) (`https://tools.ietf.org/html/rfc15`)
- **RFC 855**: *Telnet Option Specifications* (`https://tools.ietf.org/html/rfc855`)
- **RFC 4250**: *The Secure Shell (SSH) Protocol Assigned Numbers* (`https://tools.ietf.org/html/rfc4250`)
- **RFC 4251**: *The Secure Shell (SSH) Protocol Architecture* (`https://tools.ietf.org/html/rfc4251`)
- **RFC 4252**: *The Secure Shell (SSH) Authentication Protocol* (`https://tools.ietf.org/html/rfc4252`)
- **RFC 4253**: *The Secure Shell (SSH) Transport Layer Protocol* (`https://tools.ietf.org/html/rfc4253`)
- **RFC 4254**: *The Secure Shell (SSH) Connection Protocol* (`https://tools.ietf.org/html/rfc4254`)
- **RFC 4255**: *Using DNS to Securely Publish Secure Shell (SSH) Key Fingerprints* (`https://tools.ietf.org/html/rfc4255`)
- **RFC 4256**: *Generic Message Exchange Authentication for the Secure Shell Protocol (SSH)* (`https://tools.ietf.org/html/rfc4256`)

Section 4 - Odds and Ends

4

In this final section, we will study network testing and review common mistakes to watch out for in socket application programming. We will end with an exploration of the design considerations for Internet of Things connected devices.

The following chapters are in this section:

12
Network Monitoring and Security

In this chapter, we will look at common tools and techniques for **network monitoring**. These techniques can be useful to both alert us to developing problems, and to help troubleshoot existing problems. Network monitoring can be important from a network security standpoint, where it may be useful to detect, log, or even prevent network intrusions.

The following topics are covered in this chapter:

- Checking host reachability
- Displaying a connection route
- Showing open ports
- Listing open connections
- Packet sniffing
- Firewalls and packet filtering

Technical requirements

This chapter contains no C code. Instead, it focuses on useful tools and utilities.

The utilities and tools used in this chapter are either built in to your operating system or are available as free, open source software. We'll provide instructions for each tool as it is introduced.

The purpose of network monitoring

Network monitoring is a common IT term that has a broad implication. Network monitoring can refer to the practice, techniques, and tools used to provide insight into the status of a network. These techniques are used to monitor the availability and performance of networked systems and troubleshoot problems.

Some reasons you may want to practice network monitoring include the following:

- To detect the reachability of networked systems
- To measure the availability of networked systems
- To determine the performance of networked systems
- To inform decisions about network resource allocation
- To aid in troubleshooting
- To benchmark performance
- To reverse engineer a protocol
- To debug a program

In this chapter, we look at a small subset of conventional network monitoring techniques that may be useful when implementing networked programs.

When developing or deploying networked programs, it is often the case that you run into problems. When this happens, you are faced with two possibilities. The first possibility is that your program could have a bug in it. The second possibility is that your problem is caused by a network issue. The methods covered in this chapter help to identify and troubleshoot network problems.

One of the most fundamental questions you could ask is, can this system reach that system? The **Ping** utility, which is perhaps the most elementary network tool, is designed to answer just that question. Let's consider its usage next.

Testing reachability

Perhaps the most basic network monitoring tool is Ping. Ping uses the **Internet Control Message Protocol** (**ICMP**) to check whether a host is reachable. It also commonly reports the total round-trip time (latency). Ping is available as a built-in command or utility for all common operating systems.

The ICMP defines a set of special IP messages that are typically useful for diagnostic and control purposes. Ping works by using two of these messages: **echo request** and **echo reply**. The Ping utility sends an echo request ICMP message to a destination host. When that host receives the echo request, it should respond with an echo reply message.

When the echo reply is received, Ping knows that the destination host is reachable. Ping can also report the round-trip time from when the echo request was sent to when the echo reply was received. ICMP echo messages are usually small and easy to process, so this round-trip time often serves as a best-case estimate of network latency.

The Ping utility takes one argument: the hostname or address you're requesting a response from. Ping on Unix-based systems will send packets continuously. On Windows, pass in the -t flag for this behavior. Press *Ctrl +C* to stop.

The following screenshot shows using the Ping utility on example.com:

```
                    honp@ubby18: ~/Desktop

$ ping example.com
PING example.com (93.184.216.34) 56(84) bytes of data.
64 bytes from 93.184.216.34 (93.184.216.34): icmp_seq=1 ttl=128 time=49.2 ms
64 bytes from 93.184.216.34 (93.184.216.34): icmp_seq=2 ttl=128 time=355 ms
64 bytes from 93.184.216.34 (93.184.216.34): icmp_seq=3 ttl=128 time=39.3 ms
64 bytes from 93.184.216.34 (93.184.216.34): icmp_seq=4 ttl=128 time=151 ms
64 bytes from 93.184.216.34 (93.184.216.34): icmp_seq=5 ttl=128 time=34.6 ms
64 bytes from 93.184.216.34 (93.184.216.34): icmp_seq=6 ttl=128 time=66.8 ms
64 bytes from 93.184.216.34 (93.184.216.34): icmp_seq=7 ttl=128 time=65.0 ms
64 bytes from 93.184.216.34 (93.184.216.34): icmp_seq=8 ttl=128 time=62.9 ms
64 bytes from 93.184.216.34 (93.184.216.34): icmp_seq=9 ttl=128 time=186 ms
64 bytes from 93.184.216.34 (93.184.216.34): icmp_seq=10 ttl=128 time=134 ms
^C
--- example.com ping statistics ---
10 packets transmitted, 10 received, 0% packet loss, time 9018ms
rtt min/avg/max/mdev = 34.696/114.537/355.360/94.093 ms
$
```

In the preceding screenshot, you can see that several ping messages were sent, and each received a response. The round-trip time is reported for each ping message, and a summary of all sent packets is also reported. The summary includes the minimum, average, and maximum round-trip message time.

It is also possible to use ping to send larger messages. On Linux and macOS, the -s flag specifies packet size. On Windows, the -l flag is used. It is sometimes interesting to see how packet size affects latency and reliability.

The following screenshot shows pinging `example.com` with a larger 1,000-byte ping message:

```
                        honp@ubby18: ~/Desktop
$ ping -s 1000 example.com
PING example.com (93.184.216.34) 1000(1028) bytes of data.
1008 bytes from 93.184.216.34 (93.184.216.34): icmp_seq=1 ttl=128 time=39.4 ms
1008 bytes from 93.184.216.34 (93.184.216.34): icmp_seq=2 ttl=128 time=82.0 ms
1008 bytes from 93.184.216.34 (93.184.216.34): icmp_seq=3 ttl=128 time=64.4 ms
1008 bytes from 93.184.216.34 (93.184.216.34): icmp_seq=4 ttl=128 time=270 ms
1008 bytes from 93.184.216.34 (93.184.216.34): icmp_seq=5 ttl=128 time=253 ms
1008 bytes from 93.184.216.34 (93.184.216.34): icmp_seq=6 ttl=128 time=53.1 ms
1008 bytes from 93.184.216.34 (93.184.216.34): icmp_seq=7 ttl=128 time=48.9 ms
1008 bytes from 93.184.216.34 (93.184.216.34): icmp_seq=8 ttl=128 time=48.5 ms
1008 bytes from 93.184.216.34 (93.184.216.34): icmp_seq=9 ttl=128 time=50.3 ms
1008 bytes from 93.184.216.34 (93.184.216.34): icmp_seq=10 ttl=128 time=49.5 m
s
^C
--- example.com ping statistics ---
10 packets transmitted, 10 received, 0% packet loss, time 9013ms
rtt min/avg/max/mdev = 39.400/96.117/270.959/83.954 ms
$
```

If ping fails to receive an echo reply, it does not necessarily mean that the target host is unreachable. It only means that ping did not receive the expected reply. This could be because the target machine ignored the echo request. However, most systems *are* set up to respond to ping requests properly, and an echo request timeout usually means the target system is unreachable.

Sometimes, just knowing that a host is reachable is enough, but sometimes you'll want more information. It can be useful to know the exact path an IP packet takes through the network. The **traceroute** utility provides this information.

Checking a route

Although we briefly introduced traceroute in Chapter 1, *Introducing Networks and Protocols*, it is worth revisiting in more detail.

While Ping can tell us whether a network path exists between two systems, traceroute can reveal what this path actually is.

Traceroute is used with one argument: the hostname or address you want to map a route to.

On Windows, traceroute is called **tracert**. Tracert works in a very similar way to the traceroute utility found on Linux and macOS.

The following screenshot shows the traceroute utility printing the routers used to deliver data to `example.com`. The `-n` flag tells `traceroute` not to perform reverse-DNS lookups for each hop. These lookups are rarely useful and omitting them saves a bit of time and screen space:

```
                      honp@ubby18: ~/Desktop
$ traceroute -n example.com
traceroute to example.com (93.184.216.34), 30 hops max, 60 byte packets
 1  23.92.28.3  0.584 ms  0.683 ms  0.807 ms
 2  74.207.239.24  0.701 ms 74.207.239.20  0.687 ms 74.207.239.6  0.758 ms
 3  198.32.132.86  1.927 ms 74.207.239.8  0.733 ms  0.720 ms
 4  198.32.132.86  1.888 ms 152.195.80.131  3.866 ms 198.32.132.86  1.860 ms
 5  93.184.216.34  0.347 ms  0.345 ms 152.195.80.131  3.815 ms
 6  93.184.216.34  0.353 ms  0.379 ms  0.362 ms
$
```

The preceding screenshot shows that there are four or five routers (or hops) between us and the destination system at `example.com`. Traceroute also shows the round-trip time to each intermediate router.

Traceroute sends three messages to each router. This often exposes multiple network paths, and there is no guarantee that any two messages will take precisely the same path.

In the preceding example, we see that the message must first pass through `23.92.28.3`. From there, it goes to one of three different systems, which are listed. The message continues until it reaches the destination at hop five or six, depending on the exact path it takes through the network.

This illustrates an interesting point: you shouldn't assume that two consecutive packets take the same network path.

How traceroute works

To understand how traceroute works, we must understand a detail of the **Internet Protocol** (**IP**). Each IP packet header contains a field called **Time to Live** (**TTL**). TTL is the maximum number of seconds that a packet should live on the network before being discarded. This is important to keep an IP packet from simply persisting (which is, going in an endless loop) over the network.

TTL time intervals under one second are rounded up. This means that, in practice, each router that handles an IP packet decrements the TTL field by 1. Therefore, TTL is often used as a hop-count. That is to say that the TTL field simply represents the number of hops a packet can still take over the network.

The traceroute utility uses TTL to identify intermediate routers in a network. Traceroute begins by addressing a message (for example, a UDP datagram or an ICMP echo request) to the destination host. However, traceroute sets the TTL field to a value of 1. When the very first router in the connection path receives this message, it decrements TTL to zero. The router then realizes that the message has expired and discards it. A well-behaved router then sends an ICMP **Time Exceeded** message back to the original sender. Traceroute uses this **Time Exceeded** message to identify the first router in the connection.

Traceroute repeats this same process with additional messages. The second message is sent using a TTL of 2, and that message identifies the second hop in the network path. The third message is sent using a TTL of 3, and so on. Eventually, the message reaches its final destination and traceroute has mapped the entire network path.

The following diagram illustrates the method used by traceroute:

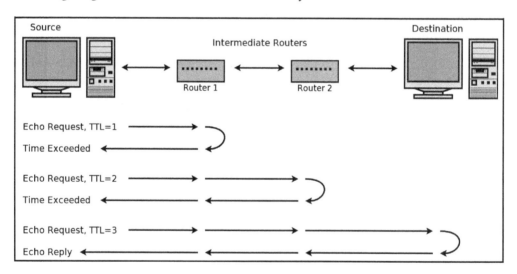

In the preceding diagram, the first message is sent with a TTL of 1. Router 1 doesn't forward this message, but instead returns an ICMP **Time Exceeded** message. The second message is sent with a TTL of 2. It makes it to the second router before timing out. The third message makes it to the destination, which replies with an **Echo Reply** message. (If this traceroute were UDP-based, it would expect to receive an ICMP **Port Unreachable** message instead.)

Not all routers return an ICMP **Time Exceeded** message, and some networks filter out these messages. In these cases, traceroute will have no way to know these routers' addresses. Traceroute prints an asterisk instead. Theoretically, if a router exists in the connection path that doesn't decrement the TTL field, then traceroute has no way of knowing that this router exists.

Now that we've covered the way ping and traceroute work, you may be wondering how they could be implemented in C code. Unfortunately, despite their simple algorithms, this is easier said than done. Keep reading to learn more.

Raw sockets

You may be interested in implementing your own network testing tools. The ping tool seems pretty simple, and it is. Unfortunately, the socket programming APIs we've been working with do not provide access at the IP level that the ICMP is built on.

The socket programming API does provide access to **raw sockets**, in theory. With raw sockets, a C program can construct the exact IP packet to send. That is, a C programmer could construct an ICMP packet from scratch and send it over the network. Raw sockets also allow for programs to receive uninterpreted packets from the network directly. In this case, the user program would be responsible for deconstructing and interpreting the ICMP packet, not the operating system.

On systems with raw socket support, getting started can be as simple as changing your `socket()` function invocation to the following:

```
socket(AF_INET, SOCK_RAW, IPPROTO_RAW);
```

However, the problem is that raw sockets aren't universally supported. It is a difficult subject to approach in a cross-platform way. Windows, in particular, has varying support for raw sockets depending on the OS version. Recent versions of Windows have virtually no support for raw sockets. For this reason, we won't cover raw sockets in any more detail here.

Now that we've covered two basic tools for network troubleshooting, let's next look at tools that inform us about our own system's relationship to the network.

Checking local connections

It is often useful to know what connections are being made on your local machine. The `netstat` command can help with that. **Netstat** is available on Linux, macOS, and Windows. Each version differs a little in the command-line options and output, but the general usage principles are the same.

I recommend running `netstat` with the −n flag. This flag prevents `netstat` from doing reverse-DNS lookups on each address and has the effect of speeding it up significantly.

On Linux, we can use the following command to show open TCP connections:

```
netstat -nt
```

The following screenshot shows the result of running this command on Linux:

```
                          honp@ubby18: ~/Desktop                         ● ▢ ⊗
$ netstat -nt
Active Internet connections (w/o servers)
Proto Recv-Q Send-Q Local Address           Foreign Address         State
tcp        0      0 192.168.182.140:57926   93.184.216.34:80        ESTABLISHED
tcp        0      0 192.168.182.140:57922   93.184.216.34:80        TIME_WAIT
tcp        0      0 192.168.182.140:37488   216.58.192.174:80       ESTABLISHED
$
```

In the preceding screenshot, you can see that `netstat` shows six columns. These columns display the protocol, the sending and receiving queue, the local address, the foreign address, and the connection state. In this example, we see that there are three connections to port 80. It is likely that this computer is loading up three web pages (as HTTP uses port 80).

On Windows, the `netstat -n -p TCP` command shows the same information, except it omits the socket queue information.

The queue information, displayed on Unix-based systems, represents how many bytes the kernel has queued up waiting for the program to read, or the number of bytes sent but not acknowledged by the foreign host. Small numbers are healthy, but if these numbers became large, it could indicate a problem with the network or a bug in the program.

It is also useful to see which program is responsible for each connection. Use the −p flag on Unix-based systems for this. On Windows, the −o flag shows PID and the −b flag shows executable names.

If you're working on a server, it is often useful to see which listening sockets are open. The `-l` flag instructs Unix-based `netstat` to show only listening sockets. The following screenshot shows listening TCP server sockets, including the program name:

In the preceding screenshot, we can see that this system is running a DNS resolver (`systemd-resolve` on port 53, IPv4 only), an SSH daemon (`sshd` on port 22), a printer service (`cupsd` on port 631), and a web server (`apache2` on port 80, IPv6 only).

On Windows, `netstat` doesn't have an easy way to display only listening sockets. Instead, you can use the `-a` flag to display everything. Filtering out only listening TCP sockets can be accomplished with the following command:

```
netstat -nao -p TCP | findstr LISTEN
```

The following screenshot shows using `netstat` on Windows to show only listening TCP sockets:

```
Administrator: C:\Windows\system32\cmd.exe                        —    □    ✕

C:\>netstat -nao -p TCP | findstr LISTEN
  TCP    0.0.0.0:135            0.0.0.0:0              LISTENING       792
  TCP    0.0.0.0:445            0.0.0.0:0              LISTENING       4
  TCP    0.0.0.0:7680           0.0.0.0:0              LISTENING       968
  TCP    0.0.0.0:49664          0.0.0.0:0              LISTENING       524
  TCP    0.0.0.0:49665          0.0.0.0:0              LISTENING       340
  TCP    0.0.0.0:49666          0.0.0.0:0              LISTENING       968
  TCP    0.0.0.0:49667          0.0.0.0:0              LISTENING       648
  TCP    0.0.0.0:49668          0.0.0.0:0              LISTENING       1580
  TCP    0.0.0.0:49670          0.0.0.0:0              LISTENING       632
  TCP    192.168.182.133:139    0.0.0.0:0              LISTENING       4

C:\>
```

With an idea of what programs are communicating from our machine, it may also be useful to see what they are communicating. Tools such as `tcpdump` and `tshark` specialize in this, and we will cover them next.

Snooping on connections

In addition to seeing which sockets are open on our computer, we can also capture the exact data being sent and received.

We have a few tooling options for this:

- **tcpdump** is a commonly used program for packet capture on Unix-based systems. It is not available on modern Windows systems, however.
- **Wireshark** is a very popular network protocol analyzer that includes a very nice GUI. Wireshark is free software, released under the GNU GPL license, and available on many platforms (including Windows, Linux, and macOS).

Included with Wireshark is Tshark, a command-line-based tool that allows us to dump and analyze network traffic. Programmers often prefer command-line tools for their simple interfaces and ease of scripting. They have the additional benefit on being usable on systems where GUIs may not be available. For these reasons, we focus on using Tshark in this section.

Tshark can be obtained from `https://www.wireshark.org`.

If you're running Linux, it is likely that your distro provides a package for Tshark. For example, on Ubuntu Linux, the following commands will install Tshark:

```
sudo apt-get update
sudo apt-get install tshark
```

Once installed, Tshark is very easy to use.

You first need to decide which network interface(s) you want to use to capture traffic. The desired interface or interfaces are passed to `tshark` with the `-i` flag. On Linux, you can listen to all interfaces by passing `-i any`. However, Windows doesn't provide the `any` interface. To listen on multiple interfaces with Windows, you need to enumerate them separately, for example, `-i 1 -i 2 -i 3`.

Tshark lists the available interfaces with the -D flag. The following screenshot shows Tshark on Windows enumerating the available network interfaces:

```
🖳 Command Prompt                                                    —    □    ✕
C:\>tshark -D
1. \Device\NPF_{3AB66933-5962-40BE-A2E8-A81CA26146DF} (Npcap Loopback Adapter)
2. \Device\NPF_{EF47CA2C-66BA-4BD5-BA92-3C7B6409238E} (Local Area Connection* 9)
3. \Device\NPF_{48F60EBF-01DD-4513-A732-43B40E47780B} (Local Area Connection* 8)
4. \Device\NPF_{07CFAB26-8E1B-41A9-89CF-4B56E74D963D} (Local Area Connection* 7)
5. \Device\NPF_{9B8B549E-60B1-4F5C-B617-EE55F26E347F} (Wire)
6. \Device\NPF_{A19D5A49-3286-4C52-ADB8-317354849FE1} (VMware Network Adapter VMnet8)
7. \Device\NPF_{BF91FBC0-ECEE-44BA-929A-F78A5C424B01} (VMware Network Adapter VMnet1)
8. \Device\NPF_{2AA46A79-14D2-4E2A-8F8C-95479B164777} (Ethernet)

C:\>
```

If you want to monitor local traffic (that is, where the communication is between two programs on the same computer), you will want to use the Loopback adapter.

Once you've identified the network interface you would like to monitor, you can start Tshark with the -i flag and begin capturing traffic. Use *Ctrl + C* to stop capturing. The following screenshot shows Tshark in use:

```
🖳 Command Promp                                                     —    □    ✕
C:\>tshark -i 5
Capturing on 'Wire'
  1   0.000000 184.29.93.254 → 192.168.50.119 TCP 60 80 → 2200 [FIN, ACK] Seq=1 Ack=1 Win=903 Len=0
  2   0.000031 192.168.50.119 → 184.29.93.254 TCP 54 2200 → 80 [ACK] Seq=1 Ack=2 Win=256 Len=0
  3   0.000073 192.168.50.119 → 184.29.93.254 TCP 54 2200 → 80 [FIN, ACK] Seq=1 Ack=2 Win=256 Len=0
  4   0.000910 192.168.50.1 → 192.168.50.119 ICMP 82 Redirect          (Redirect for host)
  5   0.061462 184.29.93.254 → 192.168.50.119 TCP 60 80 → 2200 [ACK] Seq=2 Ack=2 Win=903 Len=0
  6   0.420941 192.168.50.1 → 239.255.255.250 SSDP 303 NOTIFY * HTTP/1.1
  7   0.421612 192.168.50.1 → 239.255.255.250 SSDP 375 NOTIFY * HTTP/1.1
  8   0.422272 192.168.50.1 → 239.255.255.250 SSDP 371 NOTIFY * HTTP/1.1
  9   0.422915 192.168.50.1 → 239.255.255.250 SSDP 351 NOTIFY * HTTP/1.1
 10   0.423566 192.168.50.1 → 239.255.255.250 SSDP 383 NOTIFY * HTTP/1.1
 11   0.424233 192.168.50.1 → 239.255.255.250 SSDP 365 NOTIFY * HTTP/1.1
 12   0.424869 192.168.50.1 → 239.255.255.250 SSDP 367 NOTIFY * HTTP/1.1
 13   0.425493 192.168.50.1 → 239.255.255.250 SSDP 367 NOTIFY * HTTP/1.1
 14   0.654610 192.168.50.119 → 239.255.255.250 SSDP 179 M-SEARCH * HTTP/1.1
 15   0.656185 192.168.50.1 → 192.168.50.119 SSDP 341 HTTP/1.1 200 OK
 16   1.650339 BelkinIn_6e:5c:a8 → AsustekC_c5:79:0d ARP 60 Who has 192.168.50.119? Tell 192.168.50.1
 17   1.650348 AsustekC_c5:79:0d → BelkinIn_6e:5c:a8 ARP 42 192.168.50.119 is at 08:62:66:c5:79:0d
 18   3.654787 192.168.50.119 → 239.255.255.250 SSDP 179 M-SEARCH * HTTP/1.1
 19   3.656344 192.168.50.1 → 192.168.50.119 SSDP 341 HTTP/1.1 200 OK
```

The preceding screenshot represents only a few seconds of running Tshark on a typical Windows desktop. As you can see, there is a lot of traffic into and out of an even relatively idle system.

To cut down on the noise, we need to use a capture filter. Tshark implements a small language that allows easy specification of which packets to capture and which to ignore.

Explaining filters may be easiest by way of example.

For example, if we want to capture only traffic to or from the IP address 8.8.8.8, we will use the host 8.8.8.8 filter.

In the following screenshot, we've run Tshark with the host 8.8.8.8 filter:

```
Command Prompt                                                          —  □  ×

C:\>tshark -i 5 host 8.8.8.8
Capturing on 'Wire'
    1   0.000000 192.168.50.119 → 8.8.8.8       DNS 71 Standard query 0xabcd A example.com
    2   0.146698        8.8.8.8 → 192.168.50.119 DNS 87 Standard query response 0xabcd A example.com A 93.184.216.34
2 packets captured

C:\>
```

You can see that while Tshark was running, it captured two packets. The first packet is a DNS request sent to 8.8.8.8. Tshark informs us that this DNS request is for the A record of example.com. The second packet is the DNS response received from 8.8.8.8. Tshark shows that the DNS query response indicators that the A record for example.com is 93.184.216.34.

Tshark filters also support the Boolean operators and, or, and not. For example, to capture only traffic involving the IP addresses 8.8.8.8 and 8.8.4.4 you can use the host 8.8.8.8 filter or the host 8.8.4.4 filter.

Filtering by port number is also very useful and can be done with `port`. For example, the following screenshot shows Tshark being used to capture traffic to `93.184.216.34` on port `80`:

```
Command Prompt                                                              —  □  ×

C:\>tshark -i 5 host 93.184.216.34 and port 80
Capturing on 'Wire'
    1   0.000000 192.168.50.119 → 93.184.216.34 TCP 66 5521 → 80 [SYN] Seq=0 Win=64240 Len=0 MSS=1460 WS=256 SACK_PERM=1
    2   0.054634 93.184.216.34 → 192.168.50.119 TCP 66 80 → 5521 [SYN, ACK] Seq=0 Ack=1 Win=27800 Len=0 MSS=1370 SACK_PER
M=1 WS=32
    3   0.054705 192.168.50.119 → 93.184.216.34 TCP 54 5521 → 80 [ACK] Seq=1 Ack=1 Win=65536 Len=0
    4   0.054982 192.168.50.119 → 93.184.216.34 HTTP 145 GET / HTTP/1.1
    5   0.095022 93.184.216.34 → 192.168.50.119 TCP 60 80 → 5521 [ACK] Seq=1 Ack=92 Win=27808 Len=0
    6   0.098056 93.184.216.34 → 192.168.50.119 TCP 1424 HTTP/1.1 200 OK  [TCP segment of a reassembled PDU]
    7   0.098057 93.184.216.34 → 192.168.50.119 HTTP 300 HTTP/1.1 200 OK  (text/html)
    8   0.098057 93.184.216.34 → 192.168.50.119 TCP 60 80 → 5521 [FIN, ACK] Seq=1617 Ack=92 Win=27808 Len=0
    9   0.098140 192.168.50.119 → 93.184.216.34 TCP 54 5521 → 80 [ACK] Seq=92 Ack=1618 Win=65536 Len=0
   10   0.102372 192.168.50.119 → 93.184.216.34 TCP 54 5521 → 80 [FIN, ACK] Seq=92 Ack=1618 Win=65536 Len=0
   11   0.155505 93.184.216.34 → 192.168.50.119 TCP 60 80 → 5521 [ACK] Seq=1618 Ack=93 Win=27808 Len=0
11 packets captured

C:\>
```

In the preceding screenshot, we see that Tshark was run with the `tshark -i 5 host 93.184.216.34 and port 80` command. This has the effect of capturing all traffic on network interface `5` that is to or from `93.184.216.34` port `80`.

TCP connections are sent as a series of packets. Although Tshark reports capturing 11 packets, these are all associated with only one TCP connection.

So far, we've been using Tshark in a way that causes it to display a summary of each packet. Often this is enough, but sometimes you will want to be able to see the entire contents of a packet.

Deep packet inspection

If we pass `tshark` the `-x` flag, it displays an ASCII and hex dump of every packet captured. The following screenshot shows this usage:

```
Command Prompt                                              —    □    ×

C:\>tshark -i 5 -x host 93.184.216.34 and port 80
Capturing on 'Wire'
0000  14 91 82 6e 5c a8 08 62 66 c5 79 0d 08 00 45 00   ...n\..bf.y...E.
0010  00 34 56 4a 40 00 80 06 7b 7f c0 a8 32 77 5d b8   .4VJ@...{...2w].
0020  d8 22 58 69 00 50 a4 b0 17 bb 00 00 00 00 80 02   ."Xi.P..........
0030  fa f0 36 00 00 00 02 04 05 b4 01 03 03 08 01 01   ..6............
0040  04 02                                              ..

0000  08 62 66 c5 79 0d 00 24 1d 7d ac 29 08 00 45 58   .bf.y..$.}.)..EX
0010  00 34 00 00 40 00 3c 06 15 72 5d b8 d8 22 c0 a8   .4..@.<..r].."..
0020  32 77 00 50 58 69 8f 5b 75 d9 a4 b0 17 bc 80 12   2w.PXi.[u.......
0030  6c 98 bf 6f 00 00 02 04 05 5a 01 01 04 02 01 03   l..o.....Z......
0040  03 05                                              ..

0000  14 91 82 6e 5c a8 08 62 66 c5 79 0d 08 00 45 00   ...n\..bf.y...E.
0010  00 28 56 4b 40 00 80 06 7b 8a c0 a8 32 77 5d b8   .(VK@...{...2w].
0020  d8 22 58 69 00 50 a4 b0 17 bc 8f 5b 75 da 50 10   ."Xi.P.....[u.P.
0030  01 00 6b 7e 00 00                                  ..k~..

0000  14 91 82 6e 5c a8 08 62 66 c5 79 0d 08 00 45 00   ...n\..bf.y...E.
0010  00 83 56 4c 40 00 80 06 7b 2e c0 a8 32 77 5d b8   ..VL@...{...2w].
0020  d8 22 58 69 00 50 a4 b0 17 bc 8f 5b 75 da 50 18   ."Xi.P.....[u.P.
0030  01 00 ea fb 00 00 47 45 54 20 2f 20 48 54 54 50   ......GET / HTTP
0040  2f 31 2e 31 0d 0a 48 6f 73 74 3a 20 65 78 61 6d   /1.1..Host: exam
0050  70 6c 65 2e 63 6f 6d 3a 38 30 0d 0a 43 6f 6e 6e   ple.com:80..Conn
0060  65 63 74 69 6f 6e 3a 20 63 6c 6f 73 65 0d 0a 55   ection: close..U
0070  73 65 72 2d 41 67 65 6e 74 3a 20 68 6f 6e 70 77   ser-Agent: honpw
0080  63 20 77 65 62 5f 67 65 74 20 31 2e 30 0d 0a 0d   c web_get 1.0...
0090  0a                                                 .
```

In the preceding screenshot, you can see that entire IP packets are being dumped. In this case, we see the first three packets represent a new TCP connection's three-way handshake. The fourth packet contains an HTTP request.

This direct insight into each packet's contents isn't always convenient. It's often more practical to capture packets to a file and do the analysis later. The -w option is used with tshark to capture packets to a file. You may also want to use the -c option to limit the number of packets captured. This simple precaution protects against accidentally filling your entire hard drive with network traffic.

The following screenshot shows using Tshark to capture 50 packets to a file named capture.pcap:

```
C:\example>tshark -i 5 -w capture.pcap -c 50 tcp and port 80
Capturing on 'Wire'
50

C:\example>
```

Once the traffic is written to a file, we can use Tshark to analyze it at our leisure. Simply run tshark -r capture.pcap to get started. For text-based protocols (such as HTTP or SMTP), it is also often useful to open the capture file in a text editor for analysis.

The ultimate way to analyze captured traffic is with **Wireshark**. Wireshark allows you to load a capture file produced with tshark or tcpdump and analyze it with a very nice GUI. Wireshark is also able to understand many standard protocols.

The following screenshot shows using Wireshark to display the traffic captured from Tshark:

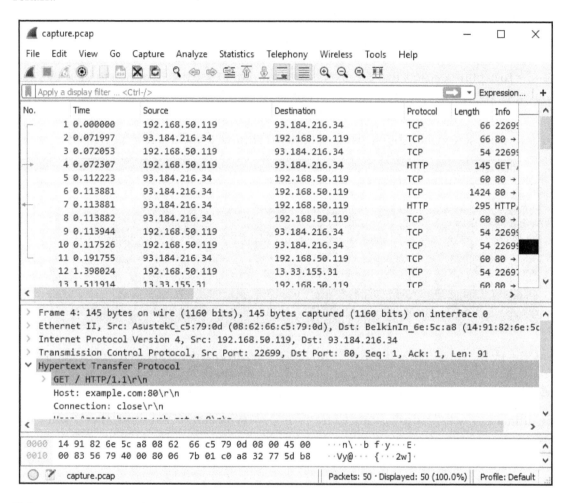

If the system you need to capture traffic on has a GUI, you can also use Wireshark to capture your traffic directly.

If you play around with `tshark` and Wireshark, you will quickly see that it is easy to inspect network protocols at a deep level. You may even find some questionable security choices made by software running on your own system.

Although we've been focusing on monitoring only network traffic on our local system, it is also possible to monitor all local network traffic.

Capturing all network traffic

`tshark` is only able to see traffic that arrives on your machine. This usually means you can only use it to monitor traffic from applications on your computer.

To capture all internet traffic on your network, you must somehow arrange for that traffic to arrive at your system, even though it usually wouldn't. There are two basic methods to do this.

The first method is by using a router that supports mirroring. This feature makes it mirror all traffic to a given Ethernet port. If you are using such a router, you can configure it to mirror all network traffic to a particular port, and then plug your computer into that port. From that point, any traffic capturing tool, such as `tcpdump` or `tshark`, can be used to record this traffic.

The second way to sniff all internet traffic is to install a hub (or a switch with port mirroring) between your router and internet modem. Hubs work by mirroring all traffic to all ports. Hubs used to be a common way to build a network. However, they have mostly been replaced by more efficient switches.

With a hub installed between your router and Internet modem, you can then connect your system directly to this hub. With that setup, you can receive all the internet traffic into or out of the network.

To monitor all network traffic (such as, even traffic between devices on the same network), you need to build your network with either a hub or a switch that supports port mirroring.

It should also be noted, that with the right modem, you can potentially capture all Wi-Fi traffic wirelessly.

From a security perspective, let this be instructive. You should never consider any network traffic to be secret. Only traffic which is secured appropriately using encryption is resistant to monitoring.

Now that we've shown several tools and techniques for testing and monitoring network traffic, let's consider the important topic of network security next.

Network security

Network security encompasses the tools, techniques, and practices to protect a network from threats. These tools include both hardware and software and can protect against a variety of threats.

Although the topic is too broad for much detail here, we will cover a few topics that you are likely to encounter.

Firewalls are one of the most common network security techniques. Firewalls act as a barrier between one network and another. As commonly used, they monitor network traffic and allow or block traffic based on a defined set of rules.

Firewalls come in two types: software and hardware. Most operating systems provide software firewalls now. Software firewalls are typically configured to deny incoming connections unless a rule is set up to allow it explicitly.

It is also possible to configure software firewalls to deny outgoing traffic by default. In this case, programs aren't allowed to establish new connections unless a specific rule is added to the firewall configuration first.

Do be careful about assuming that firewalls catch all traffic. For example, the Windows 10 firewall can be configured to deny all outgoing traffic by default, but it still lets DNS requests through. An attacker could take advantage of this by using DNS requests to exfiltrate data even though the user thinks that they are protected via the Windows firewall.

Hardware firewalls come with various capabilities. Often, they are configured to block any incoming connections that don't match predefined rules. On networks without firewalls, routers often implicitly serve this same purpose. If your router provides network address translation, then it wouldn't even know what to do with an incoming connection unless a port forwarding rule has been pre-established for it.

Although it is important to understand the basics of network security on a wide level, as C programmers, we are often more concerned with the security of our own programs. Security is not something that the C language makes easy, so let's consider application-level security in more detail now.

Application security and safety

When programming in C, special consideration must be given to security. This is because C is a low-level programming language that gives direct access to system resources. Memory management, for example, must be done manually in C, and a mistake in memory management could allow a network attacker to write and execute arbitrary code.

With C, it is vital to ensure that allocated memory buffers aren't written past the end. That is, whenever you copy data from the network into memory, you must ensure that enough memory was allocated to store the data. If your program misses this even once, it potentially allows a window for an attacker to gain control of your program.

Memory management, from a security perspective, isn't a concern with many higher-level programming languages. In many programming languages, it isn't even possible to write outside of allocated memory. Of course, these languages also can't provide the precise control of memory layout and data structures that C programmers enjoy.

Even if you are careful to manage memory perfectly, there are still many more security gotchas to be on the lookout for. When implementing any network protocol, you should never assume that the data your program receives will adhere to the protocol specification. If your program does make these assumptions, it will be open to attack by a rogue program that doesn't. These protocol bugs are a concern in any programming language, not just C. Again, though, the thing about C is that these protocol implementation errors can quickly lead to memory errors, and memory errors quickly become serious.

If you're implementing a C program intended to run as a server, you should employ a **defense in depth** approach. That is, you should set up your program in such a manner that multiple defenses must be overcome before an attacker could cause damage.

The first layer of defense is writing your program without bugs. Whenever dealing with received network data, carefully consider what would happen if the received data was not at all what you're expecting. Make sure your program does the appropriate thing.

Also, don't run your program with any more privileges than needed for its functionality. If your program doesn't need access to a sensitive set of files, make sure your operating system doesn't allow it to access those files. Never run server software as root.

If you're implementing an HTTP or HTTPS server, consider not connecting your program directly to the internet. Instead, use a reverse proxy server as the first point of contact to the internet, and have your software interface only with the proxy server. This provides an additional layer of isolation against attacks.

Finally, consider not writing networked code in C at all, if you can find an alternative. Many C servers can be rewritten as CGI programs that don't directly interact with the network at all. TCP or UDP servers can often be rewritten to use `inetd`, and thus avoid the socket programming interface altogether. If your program needs to load web pages, consider using a well-tested library, such as `libcurl`, for that purpose instead of rolling your own.

We've covered a few network testing techniques so far. When deploying these techniques on live networks, it is important to be considerate. Let's finish with a note on etiquette.

Network-testing etiquette

When network testing, it is always important to behave responsibly and ethically. Generally speaking, don't test someone else's network without their explicit permission. Doing otherwise could cause embarrassment at best, and land you in serious legal trouble at worst.

You should also be aware that some network testing techniques can set off alarms. For example, many network administrators monitor their network's load and performance characteristics. If you decide to load-test these networks without notice, you might set off automated alarms causing inconvenience.

Some other testing techniques can look like attacks. Port scanning, for example, is a useful technique where a tester tries establishing many connections on different ports. It's used to discover which ports on a system are open. However, it is a common technique used by malicious attackers to find weaknesses. Some system administrators consider port scans to be an attack, and you should never port scan a system without permission.

Summary

In this chapter, we covered a lot of ground over the broad topics of network monitoring and security. We looked at tools useful for testing the reachability of networked devices. We learned how to trace a path through the network, and how to monitor connections made on our local machine. We also discovered how to log and inspect network traffic.

We discussed network security and how it may impact the C developer. By showing how network traffic can be directly inspected, we learned first-hand the importance of encryption for communication privacy. The importance of security at the application level was also discussed.

In the next chapter, we take a closer look at how our coding practices affect program behavior. We also discuss many essential odds and ends for writing robust network applications in C.

Questions

Try these questions to test your knowledge from this chapter:

1. Which tool would you use to test the reachability of a target system?
2. Which tool lists the routers to a destination system?
3. What are raw sockets used for?
4. Which tools list the open TCP sockets on your system?
5. What is one of the biggest concerns with security for networked C programs?

The answers to these questions can be found in Appendix A, *Answers to Questions*.

Further reading

For more information about the topics covered in this chapter, please refer to the following:

- **RFC 792**: *Internet Control Message Protocol* (https://tools.ietf.org/html/rfc792)
- Wireshark (https://www.wireshark.org)

13
Socket Programming Tips and Pitfalls

This chapter builds on all the knowledge you've obtained throughout this book.

Socket programming can be complicated. There are many pitfalls to avoid and subtle programming techniques to implement. In this chapter, we consider some nuanced details of network programming that are essential for writing robust programs.

The following topics are covered in this chapter:

- Error handling and error descriptions
- TCP handshakes and orderly release
- Timeout on `connect()`
- Preventing TCP deadlocks
- TCP flow control
- Avoiding address-in-use errors
- Preventing `SIGPIPE` crashes
- Multiplexing limitations of `select()`

Technical requirements

Any modern C compiler can compile the example programs from this chapter. We recommend **MinGW** on Windows and **GCC** on Linux and macOS. See `Appendix B`, *Setting Up Your C Compiler On Windows*, `Appendix C`, *Setting Up Your C Compiler On Linux*, and `Appendix D`, *Setting Up Your C Compiler On macOS*, for compiler setup.

The code for this book can be found at: `https://github.com/codeplea/Hands-On-Network-Programming-with-C`.

From the command line, you can download the code for this chapter with the following command:

```
git clone https://github.com/codeplea/Hands-On-Network-Programming-with-C
cd Hands-On-Network-Programming-with-C/chap13
```

Each example program in this chapter runs on Windows, Linux, and macOS. When compiling on Windows, each example program will require linking to the **Winsock** library. This can be accomplished by passing the `-lws2_32` option to `gcc`.

All of the example programs in this chapter require the same header files and C macros that we developed in `Chapter` 2, *Getting to Grips with Socket APIs*. For brevity, we put these statements in a separate header file, `chap13.h`. For an explanation of these statements, please refer to `Chapter` 2, *Getting to Grips with Socket APIs*.

The first part of `chap13.h` includes the needed networking headers for each platform. The code for that is as follows:

```
/*chap13.h*/

#if defined(_WIN32)
#ifndef _WIN32_WINNT
#define _WIN32_WINNT 0x0600
#endif
#include <winsock2.h>
#include <ws2tcpip.h>
#pragma comment(lib, "ws2_32.lib")

#else
#include <sys/types.h>
#include <sys/socket.h>
#include <netinet/in.h>
#include <arpa/inet.h>
#include <netdb.h>
#include <unistd.h>
#include <errno.h>
#include <fcntl.h>

#endif
```

We also define some macros to make writing portable code easier, and we include the additional headers that our programs need:

```
/*chap13.h continued*/

#if defined(_WIN32)
#define ISVALIDSOCKET(s) ((s) != INVALID_SOCKET)
#define CLOSESOCKET(s) closesocket(s)
#define GETSOCKETERRNO() (WSAGetLastError())

#else
#define ISVALIDSOCKET(s) ((s) >= 0)
#define CLOSESOCKET(s) close(s)
#define SOCKET int
#define GETSOCKETERRNO() (errno)
#endif


#include <stdio.h>
#include <stdlib.h>
#include <string.h>
```

That concludes `chap13.h`.

Error handling

Error handling can be a problematic topic in C as it does not "hold the programmer's hand". Any memory or resources allocated must be manually released, and this can be tricky to get exactly right in every situation.

When a networked program encounters an error or unexpected situation, the normal program flow is interrupted. This is made doubly difficult when designing a multiplexed system that handles many connections concurrently.

The example programs in the book take a shortcut to error handling. Almost all of them simply terminate after an error is detected. While this is sometimes a valid strategy in real-world programs, real-world programs usually need more complicated error recovery.

Sometimes, you can get away with merely having your client program terminate after encountering an error. This behavior is often the correct response for simple command-line utilities. At other times, you may need to have your program automatically try again.

Event-driven programming can provide the technique needed to simplify this logic a bit. Mainly, your program is structured so that a data structure is allocated to store information about each connection. Your program uses a main loop that checks for events, such as a readable or writable socket, and then handles those events. When structuring your program in this way, it is often easier to flag a connection as needing an action, rather than calling a function to process that action immediately.

With careful design, errors can be handled as a simple matter of course, instead of as exceptions to the normal program flow.

Ultimately, error handling is a very specialized process, and careful care needs to be taken to consider application requirements. What's appropriate for one system is not necessarily correct for another.

In any case, a robust program design dictates that you carefully consider how to handle errors. Many programmers focus only on the happy path. That is, they take care to design the program flow based on the assumption that everything goes correctly. For robust programs, this is a mistake. It is equally important to consider the program flow in cases where everything goes wrong.

Throughout the rest of this chapter, we touch on places where network programming can go wrong. Network programming can be subtle, and many of these failure modes are surprising. However, with proper consideration, they are all capable of being handled.

Before diving into all the weird ways a connection can fail, let's first focus on making error logging a bit easier. In this book, so far, we've been dealing with numeric error codes. It is often more useful to obtain a text description of an error. We look at a method for this next.

Obtaining error descriptions

In Chapter 2, *Getting to Grips with Socket APIs,* we developed the GETSOCKETERRNO() macro as a cross-platform way to obtain the error code after a failed system call.

The GETSOCKETERRNO() macro is repeated here for your convenience:

```
#if defined(_WIN32)
#define GETSOCKETERRNO() (WSAGetLastError())
#else
#define GETSOCKETERRNO() (errno)
#endif
```

The preceding code has served us well throughout this book. It has the advantage of being short and simple.

In a real-world program, you may want to display a text-based error message in addition to the error code. Windows and Unix-based systems both provide functions for this purpose.

We can build a simple function to return the last error message as a C string. The code for this function is the following:

```
/*error_text.c excerpt*/

const char *get_error_text() {

#if defined(_WIN32)

    static char message[256] = {0};
    FormatMessage(
        FORMAT_MESSAGE_FROM_SYSTEM|FORMAT_MESSAGE_IGNORE_INSERTS,
        0, WSAGetLastError(), 0, message, 256, 0);
    char *nl = strrchr(message, '\n');
    if (nl) *nl = 0;
    return message;

#else
    return strerror(errno);
#endif

}
```

The preceding function formats the error as text using `FormatMessage()` on Windows and `strerror()` on other operating systems.

Unix-based systems provide the `strerror()` function. This function takes the error code as its only parameter, and it returns a pointer to an error message string.

Getting an error code description on Windows is a bit more involved. We use the `FormatMessage()` function to obtain the text description. This function has many options, but the parameters used in the preceding code snippet work well for our purposes. Note that Windows error descriptions are generally returned ending with a newline. Our function uses `strrchr()` to find the last line-feed character and truncate the description at that point.

This chapter's code includes a program called `error_text.c` that demonstrates this method. This program calls the `socket()` function with invalid parameters, and then uses `get_error_text()` to display the error message:

```
/*error_text.c excerpt*/

        printf("Calling socket() with invalid parameters.\n");
        socket(0, 0, 0);
        printf("Last error was: %s\n", get_error_text());
```

Note that error codes and descriptions vary greatly between operating systems. The next two screenshots show the error message displayed by this program on both Windows and Linux.

The following screenshot shows `error_text` running on Windows:

The next screenshot shows `error_text` running on an Ubuntu Linux desktop:

As you can see from the preceding two screenshots, different operating systems don't often report errors in the same way.

Now that we have a better way to investigate errors, let's move on to consider some ways that **TCP sockets** can fail.

TCP socket tips

The **Transmission Control Protocol** (**TCP**) is a fantastic protocol, and TCP sockets provide a beautiful abstraction. They present discrete packets on an unreliable network as a reliable, continuous stream of data. To the programmer, sending and receiving data from a peer anywhere in the world is made nearly as easy as reading and writing to a file.

TCP works very well to hide network shortcomings. When a flaky network drops a few packets, TCP faithfully sorts out the mess and retransmits as needed. The application using TCP receives the data in perfect order. The application doesn't even know there was a network problem, and it certainly doesn't need to address the problem.

With this abstraction, like all abstractions, comes some inherent risk. TCP tries very hard to make networks look reliable. It usually succeeds, but sometimes, abstractions leak. What happens if your network cable is cut? What happens if the application you are connected to crashes? TCP isn't magic. It can't fix these problems.

Of course, it's evident that the abstraction must break when faced with severe problems such as a total network outage. However, sometimes, more subtle problems can arise from details thought to be abstracted away. For example, what happens when you try to send a lot of data, but the peer that you are connected to isn't reading it? (Answer: the data gets backed up.)

In this section, we look at TCP in a little more detail. We're especially interested in the behavior of TCP sockets in these edge cases.

A TCP connection lifespan can be divided into three distinct phases. They are as follows:

- The setup phase
- The data-transfer phase
- The tear-down phase

Problems can arise in each step.

In the setup phase, we have to consider what happens if the target system doesn't respond. By default, `connect()` sometimes waits a long time, attempting to establish a TCP connection. Sometimes, that is what you want, but often it isn't.

For the data-transfer phase, we must be careful to prevent deadlocks. An awareness of TCP congestion control mechanisms can also help us to prevent degenerate cases where our connection becomes slow or uses a lot more bandwidth than necessary.

Finally, knowing the details of the tear-down phase helps us to ensure that we haven't lost data at the end of a connection. Details of how sockets are terminated can also cause operating systems to hold on to half-dead connections long after they've disconnected. These lingering sockets can prevent new programs from binding to their local ports.

Let's begin with some information about the three-way handshake that establishes TCP connections and how to timeout a connect() call.

Timeout on connect()

Usually, when we call connect() on a TCP socket, connect() blocks until the connection is established.

The following diagram illustrates the TCP three-way handshake that establishes a typical TCP connection and how it relates to a standard, blocking connect() call:

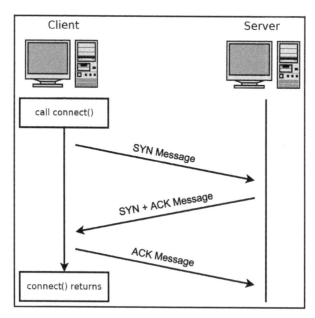

The standard TCP three-way handshake consists of three parts. First, the **Client** sends a **Synchronize (SYN)** message to the **Server**. Then the **Server** responds with an **SYN Message** of its own, combined with an **Acknowledged (ACK)** message of the **Client**'s **SYN Message**. The **Client** then responds with an acknowledgment of the **Server**'s **SYN Message**. The connection is then open and ready for data transmission.

When the `connect()` function is called on the **Client** side, the first **SYN Message** is sent, and the `connect()` function blocks until the **SYN+ACK Message** is received from the **Server**. After the **SYN+ACK Message** is received, `connect()` enqueues the final **ACK Message** and returns.

This means that `connect()` blocks for at least one round-trip network time. That is, it blocks from the time that its **SYN Message** is sent to the time that the **SYN+ACK Message** is received. While one round-trip network time is the best-case scenario, in the worst case, it could block for much longer. Consider what happens when an overloaded **Server** receives an **SYN Message**. The **Server** could take some time to reply with the **SYN+ACK Message**.

If `connect()` cannot establish a connection successfully (that is, **SYN+ACK Message** is never received), then the `connect()` call eventually times out. This timeout period is controlled by the operating system. The exact timeout period varies, but 20 seconds is about typical.

There is no standard way to extend the timeout period of `connect()`, but you can always call `connect()` again if you want to keep trying.

There are a few ways to make `connect()` timeout early. One way is to use multiple processes and kill the child process if it doesn't connect in time. Another way is to use `SIGALARM` in Unix-based systems.

A cross-platform `connect()` timeout can be achieved by using `select()`. Recall from `Chapter 3`, *An In-Depth Overview of TCP Connections*, that `select()` allows us to wait on a socket operation with a specified timeout.

`select()` also has the additional benefit of allowing your program to do useful work while waiting for the TCP connection to be established. That is, `select()` can be used to wait on multiple `connect()` calls, and other socket events besides. It can work well for a client that needs to connect to several servers in parallel.

Using `select()` to timeout a `connect()` call involves a few steps. They are as follows:

1. Set the socket to non-blocking operation. This is done using `fcntl(O_NONBLOCK)` on Unix-based systems and `ioctlsocket(FIONBIO)` on Windows.
2. Call `connect()`. This call returns immediately, provided that *step 1* was successful.
3. Check the return code from `connect()`. A return value of zero indicates that the connection was successful, which probably indicates that non-blocking mode was set incorrectly. A non-zero return value from `connect()` means we should check the error code (that is, `WSAGetLastError()` on Windows and `errno` on other platforms). An error code of `EINPROGRESS` (`WSAEWOULDBLOCK` on Windows) indicates that the TCP connection is in progress. Any other value indicates an actual error.
4. Set up and call `select()` with the desired timeout.
5. Set the socket back to blocking mode.
6. Check to see whether the socket connected successfully.

Step 1, setting the socket to non-blocking mode, can be accomplished with the following code:

```
#if defined(_WIN32)
    unsigned long nonblock = 1;
    ioctlsocket(socket_peer, FIONBIO, &nonblock);
#else
    int flags;
    flags = fcntl(socket_peer, F_GETFL, 0);
    fcntl(socket_peer, F_SETFL, flags | O_NONBLOCK);
#endif
```

The preceding code works a bit differently depending on whether it is running on Windows. On Windows, the `ioctlsocket()` function is used with the `FIONBIO` flag to indicate non-blocking socket operation. On non-Windows systems, the `fcntl()` function is used to set the `O_NONBLOCK` flag for the same purpose.

In *step 2* and *step 3*, the call to `connect()` is done normally. The only difference is that you should expect an error code of `EINPROGRESS` on Unix-based systems and `WSAEWOULDBLOCK` on Windows.

In *step 4*, the setup for `select()` is straightforward. The `select()` function is used in the same way as described in previous chapters. For your convenience, the following code shows one way to use `select()` for this purpose:

```
fd_set set;
FD_ZERO(&set);
FD_SET(socket_peer, &set);

struct timeval timeout;
timeout.tv_sec = 5; timeout.tv_usec = 0;
select(socket_peer+1, 0, &set, 0, &timeout);
```

Notice in the preceding code that we set a timeout of five seconds. Therefore, this `select()` call returns after either the connection is established, the connection has an error, or 5 seconds have elapsed.

In *step 5*, setting the socket back to non-blocking mode is accomplished with the following code:

```
#if defined(_WIN32)
    nonblock = 0;
    ioctlsocket(socket_peer, FIONBIO, &nonblock);
#else
    fcntl(socket_peer, F_SETFL, flags);
#endif
```

In *step 6*, we are looking to see whether the call to `select()` timed out, returned early from an error, or returned early because our socket has successfully connected.

Surprisingly, there is no easy, robust, cross-platform way to check whether the socket is connected at this point. My advice is to simply assume that any socket marked by `select()` as writable has connected successfully. Just try to use the socket. Most TCP client programs will want to call `send()` after connecting, anyway. The return value from this first `send()` call indicates whether you have a problem.

If you really do want to try and determine the socket state without resorting to send(), you should be aware of some differences in how select() signals in this situation. On Unix-based systems, select() signals a socket as writable once the connection is established. If an error has occurred, select() signals the socket as both writable and readable. However, if the socket has connected successfully and data has arrived from the remote peer, this also produces both the readable and writable situation. In that case, the getsockopt() function can be used to determine whether an error has occurred. On Windows, select() marks a socket as excepted if an error occurred.

Please refer to connect_timeout.c in this chapter's code repository for a working example of the connect() timeout method using select(). An additional example, connect_blocking.c, is also included for comparison.

Once a new connection is established, our concern moves to preventing data-transfer problems. In the worst case, our program could get deadlocked with its peer, preventing any data transfer. We'll consider this in more detail next.

TCP flow control and avoiding deadlock

When designing application protocols and writing network code, we need to be careful to prevent a **deadlock** state. A deadlock is when both sides on a connection are waiting for the other side to do something. The worst-case scenario is when both sides end up waiting indefinitely.

A trivial example of a deadlock is if both the client and server call recv() immediately after the connection is established. In that case, both sides wait forever for data that is never going to come.

A less obvious deadlock situation can happen if both parties try to send data at the same time. Before we can consider this situation, we must first understand a few more details of how TCP connections operate.

When data is sent over a TCP connection, this data is broken up into segments. A few segments are sent immediately, but additional segments aren't sent over the network until the first few segments are acknowledged as being received by the connected peer. This is part of TCP's **flow-control** scheme, and it helps to prevent a sender from transmitting data faster than a receiver can handle.

Consider the following diagram:

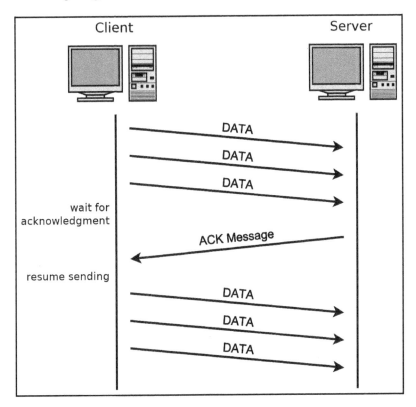

In the preceding diagram, the **Client** sends three TCP segments of data to the **Server**. The **Client** has additional **DATA** ready to send, but it must wait until the already-sent data is acknowledged. Once the **ACK Message** is received, the **Client** resumes sending its remaining **DATA**.

This is the TCP flow-control mechanism that ensures that the sender isn't transmitting faster than the receiver can handle.

Now, keeping in mind that a TCP socket can send only a limited amount of data before requiring acknowledgment of receipt, imagine what happens if both parties to a TCP connection try to send a bunch of data at the same time. In this case, both parties send the first few TCP segments. They both then wait until their peer acknowledges receipt before sending more. However, if neither party is reading data, then neither party acknowledges receiving data. This is a deadlock state. Both parties are stuck waiting forever.

Many application protocols prevent this problem by design. These protocols naturally alternate between sending and receiving data. For example, in HTTP, the client sends a request, and then the server sends a reply. The server only starts sending data after the client has finished sending.

However, TCP is a full-duplex protocol. Applications that do need to send data in both directions simultaneously should take advantage of TCP's ability to do so.

As a motivating example, imagine implementing a file-transfer program where both peers to a TCP connection are sending large parts of a file at the same time. How do we prevent the deadlock condition?

The solution to this is straightforward. Both sides should alternate calls to send() with calls to recv(). The liberal use of select() will help us do this efficiently.

Recall that select() indicates which sockets are ready to be read from and which sockets are ready to be written to. The send() function should be called only when you know that a socket is ready to be written to. Otherwise, you risk that send() may block. In the worst case, send() will block indefinitely.

Thus, one procedure to send a large amount of data is as follows:

1. Call send() with your remaining data.
2. The return value of send() indicates how many bytes were actually consumed by send(). If fewer bytes were sent than you intended, then your next call to send() should be used to transmit the remainder.
3. Call select() with your socket in both the read and write sets.
4. If select() indicates that the socket is ready to be read from, call recv() on it and handle the received data as needed.
5. If select() indicates that the socket is ready to write to again, go to *step 1* and call send() with the remaining data to be sent.

The important point is that calls to send() are interspersed with calls to recv(). In this way, we can be sure that no data is lost, and this deadlock condition does not occur.

This method also neatly extends to applications with many open sockets. Each socket is added to the select() call, and ready sockets are serviced as needed. Your application will need to keep track of which data is remaining to be sent for each connection.

It should also be noted that setting sockets to a non-blocking mode can simplify your program's logic in some cases. Even with non-blocking sockets, `select()` can still be used as a central blocking point to wait for socket events.

Two files are included with this chapter's code repository that can help to demonstrate the deadlock state and how `select()` can be used to prevent it. The first file, `server_ignore.c`, implements a simple TCP server that accepts connections and then ignores them. The second file, `big_send.c`, initiates a TCP connection and then attempts to send lots of data. By using the `big_send` program to connect to the `server_ignore` program, you can investigate the blocking behavior of `send()` for yourself.

Deadlocks represent only one way a TCP connection can unexpectedly fail. While deadlocks can be very difficult to diagnose, they are preventable with careful programming. Besides the risk for deadlock, TCP also presents other data transfer pitfalls. Let's consider another common performance problem next.

Congestion control

As we've just seen, TCP implements flow control to prevent a sender from overwhelming a receiver. This flow control works by allowing only a limited number of TCP segments to be sent before requiring an acknowledgment of receipt.

TCP also implements **congestion control** methods as part of a network **congestion-avoidance** scheme. While flow control is vital to prevent overwhelming the receiver, congestion control is essential to prevent overwhelming the network.

One way TCP congestion control works is by allowing only a limited amount of data to be sent before pausing to wait for an acknowledgment of receipt. This data limit is decreased when network congestion is detected. In this way, TCP doesn't try putting more data over the network than the network can handle.

Another way TCP implements congestion control is through the **TCP slow start algorithm**. This method provides a way for TCP to ramp-up a connection to its full potential, instead of immediately dumping a lot of data on the network all at once.

It works like this—when a new TCP connection is established, only a minimal amount of data is allowed to be sent unacknowledged. When this data is acknowledged, the limit is increased. Each time a new acknowledgment is received, the limit is increased further until packet loss happens or the limit reaches the desired maximum.

The following diagram shows a TCP slow-start in action:

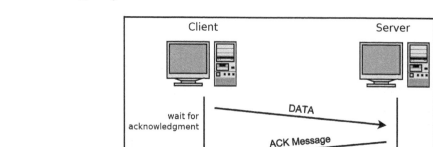

In the preceding diagram, you can see that the **Client** starts by sending only a little data. Once that data is acknowledged, the **Client** is willing to send a larger amount of data before requiring another acknowledgment. Once that acknowledgment is received, the **Client** increases its limit again, and so on.

The slow-start algorithm can cause problems for short-lived connections. In practice, if a connection needs to send only a small amount of data, that connection won't ever reach its full potential. This has caused many protocols to be designed around keeping connections open for longer. For example, it used to be common for an HTTP connection to transmit only one resource. Now it is far more common for an HTTP connection to be held open for additional resources, one after another. This connection reuse avoids the overhead of the TCP three-way handshake and slow start.

In addition to avoiding congestion, TCP also provides methods to increase bandwidth efficiency.

The Nagle algorithm

One technique used by TCP increase efficiency is the **Nagle algorithm**. The Nagle algorithm works to make the sender pool small amounts of data together until it has enough to justify sending.

Consider sending just one byte of data over a TCP connection. Each TCP segment uses 20 bytes to transmit TCP bookkeeping. An additional 20 bytes are needed for the IPv4 header. So, this 1 byte of application data becomes 41 bytes on the network. That's an overhead of 4,000%, and we are not even counting the overhead from lower layers (for example, the Ethernet frame overhead) yet!

The Nagle algorithm states that only one small, unacknowledged TCP segment may be outstanding at any given time. A small segment is considered any segment less than the **Maximum Segment Size (MSS)**.

Let's see how this applies to a program doing small writes. Consider the following code called on a connected, but otherwise idle, TCP socket:

```
send(my_socket, "a", 1, 0);
send(my_socket, "b", 1, 0);
```

After the first `send()` call, the a data is packed into a TCP message and sent off, along with its 40 bytes of TCP and IPv4 overhead.

The second `send()` call returns immediately, but the b data isn't actually sent immediately. The Nagle algorithm causes the b data to be queued-up by the operating system. It won't be sent until either the first TCP message is acknowledged or `send()` is called again with enough additional data to fill up an entire max-size TCP segment.

For both a and b to be received by the recipient, it will take the duration of one round-trip network time, plus an additional one-way network time.

We can easily get this 1.5 round-trip network time down to 0.5 round-trip network time by just using the following code:

```
send(my_socket, "ab", 2, 0);
```

For this reason, you should always prefer doing one large write to `send()` instead of many small ones, whenever possible. Doing one large write allows ab to be sent in the same TCP message, thereby sidestepping the Nagle algorithm altogether.

In some applications, you really do need to send a small packet followed by another small packet immediately after. For example, in a real-time multiplayer video game, you can't queue up player commands; they must be sent continuously. In these cases, it makes sense to disable the Nagle algorithm for reduced latency, at the expense of decreased bandwidth efficiency.

Disabling the Nagle algorithm can be done using the `setsockopt()` function. The following code shows this method in action:

```
int yes = 1;
if (setsockopt(my_socket, IPPROTO_TCP, TCP_NODELAY,
        (void*)&yes, sizeof(yes)) < 0) {
    fprintf(stderr, "setsockopt() failed. (%d)\n", GETSOCKETERRNO());
}
```

Be sure to consider all your options before disabling Nagle. When faced with a poorly performing network program, some programmers will disable the Nagle algorithm as a first step. In reality, the decision to disable the Nagle algorithm should be approached cautiously. Disabling the Nagle algorithm in real-time applications often makes sense. Disabling it in other contexts rarely does.

For example, imagine that you've implemented an HTTP client. It seems a bit sluggish, and so you try disabling the Nagle algorithm. You do that and find that it runs much faster now. However, by disabling the Nagle algorithm, you've increased network overhead. You could have gotten the same improvement by simply pooling together your `send()` calls.

If you're implementing a real-time algorithm that does need to send small time-critical packets, using `TCP_NODELAY` may still not be the right method for you. TCP can introduce delays in many other ways. For example, if one TCP packet is lost over the network, no further data can be delivered until that packet is retransmitted. This can have the effect of delaying many packets because of one hold-up.

Many real-time applications prefer using UDP over TCP. Each UDP packet is entirely independent of any other packets sent before or after. Of course, the trade-off is that there is a lesser guarantee of reliable delivery; messages may arrive in a different order than they were set, and some messages may arrive twice. Nonetheless, many applications can tolerate this. Real-time video streaming, for example, can use UDP, where each packet stores a very short, time-stamped part of the video. If a packet is lost, there is no need to retransmit; the video stutters for a moment and resumes when the next packet arrives. Packets received late, or out of order, are safely ignored.

Although the Nagle algorithm often works well to improve network utilization, not understanding how it works can lead to problems. In addition to the Nagle algorithm, TCP implements many other methods to limit the needless waste of network resources. Sometimes, these other methods work poorly with one another. The **delayed ACK** is one such method that can work badly with the Nagle algorithm.

Delayed acknowledgment

We've seen that many client-server protocols work by having the client send a request and then the server send a response. We've also seen that when a TCP peer reads data off the network, it sends an acknowledgment to let the sender know that the data was received successfully.

A typical client-server interchange might, therefore, look like the following:

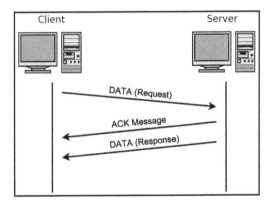

In the preceding diagram, the **Client** first sends a request to the **Server**. The **Server** reads this request, and a TCP **ACK Message** is sent back to the **Client**. The **Server** then processes the request data and replies with its response.

Some TCP stacks implement a **delayed acknowledgment** method to reduce network congestion. This technique works by delaying the acknowledgment of received data. The hope is that the receiver is going to send a response very soon anyway, and that the acknowledgment can piggyback on this response. When it works, which is often, it conserves bandwidth.

If the receiver doesn't send a reply, the acknowledgment is sent after a short delay; 200 milliseconds is typical.

If the server from before implements delayed acknowledgment, the client-server interchange might look like the following:

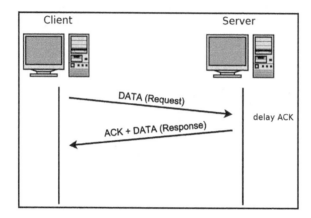

This is when delayed acknowledgment works well.

Now, consider combining the Nagle algorithm with delayed acknowledgment. If the client transmits its request in two small messages, then the sending channel is blocked for not only the round-trip time. It is also blocked for the additional acknowledgment delay time.

This is illustrated in the following diagram:

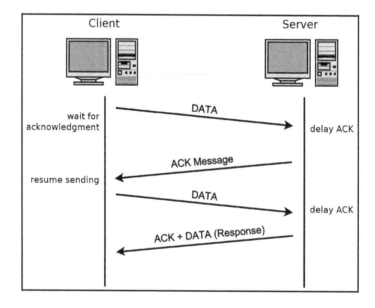

In the preceding diagram, we see that the **Client** sent the first part of its request in a small packet. The Nagle algorithm prevents it from sending the second part of its request until it receives an acknowledgment from the **Server**. Meanwhile, the **Server** receives the request, but it delays acknowledgment in the hope that it can piggyback the **ACK Message** on the reply. The **Server** processes the first part of the request and sees that it doesn't have the full request yet, so it cannot send a reply. After the delay period elapses, the **Server** does eventually send an **ACK Message**. The **Client** receives this **ACK Message** and sends the rest of the reply. The **Server** replies with its response.

In this degenerate case, the interaction of the Nagle algorithm and the delayed acknowledgment technique cased the **Client-Server** interaction to take two full round-trip network times plus the delayed acknowledgment time (which could itself be many round-trip times).

Some programmers jump in these situations to disable the Nagle algorithm. Sometimes that is needed, but often it is the wrong solution.

In our example, merely passing larger data buffers to `send()` completely solves the degenerate interaction. Passing the entire request to `send()` in one call reduces the transaction time from two round-trips plus delay to one round-trip and no delay.

My advice is to prefer calling `send()` with one large write instead of multiple small writes, whenever possible. Of course, if you're implementing a real-time application with TCP, then you can't pool `send()` calls. In that case, disabling the Nagle algorithm can be the correct call.

For the sake of completeness, it should be noted that a delayed ACK can usually be disabled. This is done by passing `TCP_QUICKACK` to `setsockopt()` on systems that support it. Again, this is not usually needed.

Now that we've reviewed a few hidden problems that can crop up with active TCP connections, it's time to move on to connection teardown.

Connection tear-down

The way a TCP connection transitions from an established connection to a closed one is nuanced. Let's consider this in more detail.

TCP connections are **full-duplex**. This means that the data being sent is independent of the data being received. Data is sent and received simultaneously. This also implies that the connection must be closed by both sides before it is truly disconnected.

To close a TCP connection, each side sends a **Finish** (**FIN**) message and receives an ACK message from their peer.

The exact tear-down process, from the perspective of each peer, depends on whether it sent a FIN first, or received a FIN first. There are three basic connection tear-down cases. They are as follows:

1. You initiate the tear-down by sending the first FIN message
2. You receive a FIN message from your connected peer
3. You and your peer send FIN messages simultaneously

In case 3, where both sides send a FIN message simultaneously, each side thinks that it is in case 1. That is, each side thinks that it has sent the first FIN message, and each side tears down its socket as in case 1. In practice, this is pretty rare, but certainly possible.

When a TCP socket is open for full-duplex communication, it is said to be in the ESTABLISHED state. The closing initiator sends a FIN message to its peer. The peer replies with an ACK. At this point, the connection is only half closed. The initiator can no longer send data, but it can still receive data. The peer has the option to continue to send more data to the closing initiator. When the peer is ready to finish closing the connection, it sends its own FIN message. The initiator then responds with the final ACK message, and the connection is fully closed.

The TCP connection state transitions on the initiator are ESTABLISHED, FIN-WAIT-1, FIN-WAIT-2, TIME-WAIT, and CLOSED. The TCP connection state transitions on the receiving peer are ESTABLISHED, CLOSE-WAIT, LAST-ACK, and CLOSED.

The following diagram illustrations the normal TCP four-way closing handshake:

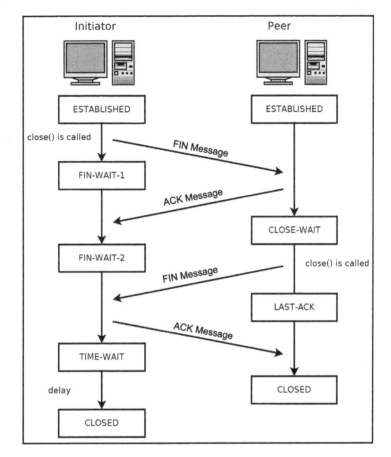

It is sometimes possible for the **Peer** to combine its **ACK Message** and **FIN Message** into one message. In that case, the connection can be torn down with only three messages, instead of four.

In the case where both sides initiate the tear-down simultaneously, both sides follow the state transition of the **Initiator**. The messages sent and received are the same.

Networks are inherently unreliable, so there is a chance that the final **ACK Message** sent by the **Initiator** will be lost. In this case, the **Peer**, having not received an **ACK Message**, resends its **FIN Message**. If the **Initiator** had completely **CLOSED** its socket after sending the final **ACK Message**, then it would be impossible to reply to this resent **FIN Message**. For this reason, the **Initiator** enters a TIME-WAIT state after sending the last **ACK Message**. During this TIME-WAIT state, it responds to any retransmitted **FIN Message** from the **Peer** with an **ACK Message**. After a delay, the **Initiator** leaves the TIME-WAIT state and fully closes its socket.

The TIME-WAIT delay is usually on the order of one minute, but it could be configured for much longer.

In this book, we've used only the close() function (closesocket() on Windows) to disconnect a socket. This function, although simple to use, has the disadvantage of always fully closing a socket. That is, no data can be sent or received on a socket called with close(). The TCP teardown handshake does allow for data to be received after a FIN message has been sent. Let's next consider how to do this programmatically.

The shutdown() function

As we've just seen, TCP connections are torn down in two steps. The first one side sends a FIN message, and then the other side does. However, each side is allowed to continue to send data until it has sent its own FIN message.

We've used the close() function (closesocket() on Windows) to disconnect sockets because of its simplicity. The close() function, however, closes both sides of a socket. If you use close() in your application, and the remote peer tries to send more data, it will cause an error. Your system will then transmit a **Reset** (**RST**) message to indicate to the peer that the connection was not closed in an orderly manner.

If you want to close your sending channel, but still leave the option for receiving more data, you should use the shutdown() function instead. The shutdown() function takes two parameters. The first parameter is a socket, and the second is an int indicating how to shut down the socket.

In theory, shutdown() supports three options—closing the sending side of a connection, closing the receiving side, and closing both sides. However, the TCP protocol itself doesn't reflect these options, and it is rarely useful to use shutdown() for closing the receiving side.

There is a small portability issue about `shutdown()` functions parameters. Under windows, you want to call it with `SD_SEND`. On other systems, you should use `SHUT_WR`. Both values are defined as `1`, so you can also call it that way.

The code to shutdown the sending channel of a socket in a cross-platform manner is as follows:

```
if (shutdown(my_socket, 1) /* 1 = SHUT_WR, SD_SEND */) {
    fprintf(stderr, "shutdown() failed. (%d)\n", GETSOCKETERRNO());
}
```

This use of `shutdown()` causes the TCP FIN message to be transmitted after the transmission queued is emptied.

You may wonder, if you're receiving data from a peer, and `recv()` returns `0`, how you know whether your peer has called `shutdown()` or `close()`? Unfortunately, you can't know, except by prior agreement. If they have used `shutdown()` only to close their sending data channel, then they are still receptive to additional data. If they instead used `close()`, additional data will trigger an error state.

Although half-closed connections have their uses, it is often easier to use an application protocol that clearly indicates the end of the transaction. For example, consider the HTTP protocol covered in `Chapter 6`, *Building a Simple Web Client*. With the HTTP protocol, the client indicates the end of its request with a blank line. The server knows it has the full request when it sees this blank line. The server then specifies how much data it will be sending with the `Content-Length` header. Once the client has received that much data, it knows that it hasn't missed anything. The client can then call `close()` and be confident that the server won't be sending additional data.

In many applications, knowing whether the shutdown was orderly isn't always useful. Consider the chat room program (`tcp_serve_chat.c`) from `Chapter 3`, *An In-Depth Overview of TCP Connections*. This program has no real application protocol. That program simply sends messages from one client to every other client. When a client decides to disconnect, it isn't important that it continues to receive data from the server. Guaranteeing an orderly TCP release would provide no benefit.

So, when is `shutdown()` useful? Basically, a TCP orderly release is useful when the application protocol doesn't have a way to signal that it has finished sending data, and your application isn't tolerant of missed data. In that case, `shutdown()` is a useful signal.

Please note that if you're using threading or forking, there are additional differences to the behavior of `close()` versus `shutdown()` that must be considered. When `shutdown()` is called, it always affects the socket. The `close()` function, by contrast, has no effect if additional processes also hold handles to the socket.

Finally, note that `close()` must still eventually be called on a socket closed with `shutdown()` in order to release associated system resources.

Another issue that comes up around the TCP tear-down procedure is the long delay for the side that initiated the close to remain in the `TIME-WAIT` state. This can sometimes cause problems for TCP servers. Let's look at that next.

Preventing address-in-use errors

If you do TCP server programming for very long, you will eventually run into the following scenario—your TCP server has one or more open connections, and then you terminate it (or it crashes). You restart the server, but the call to `bind()` fails with an `EADDRINUSE` (`WSAEADDRINUSE` on Windows) error.

When this happens, you can wait a few moments, try it again, and it works. What's going on here?

Essentially, when an application initiates a TCP socket close (or causes the disconnection by crashing), that socket goes into the `TIME-WAIT` state. The operating system continues to keep track of this socket for some time, potentially minutes.

An example program, `server_noreuse.c`, is included in this chapter's code repo. You can reproduce this address-in-use problem by running it, accepting a connection, and then terminating `server_noreuse`. To reproduce the problem, it is vital that the server is the one to terminate the open connection, not the client.

If you immediately start `server_noreuse` again, you will see the `bind()` error.

The following screenshot shows this on a Linux desktop:

```
honp@ubby18: ~                                    ⊖ ⊡ ⊗
honp@ubby18:~$ gcc server_noreuse.c -o server_noreuse
honp@ubby18:~$ ./server_noreuse
Configuring local address...
Creating socket...
Binding socket to local address...
Listening...
Waiting for connection...
Client is connected.
Waiting for connection...
^C
honp@ubby18:~$ ./server_noreuse
Configuring local address...
Creating socket...
Binding socket to local address...
bind() failed. (98)
honp@ubby18:~$
```

You can use the netstat command to see these half-dead connections that are preventing our server from starting. The following command shows which connections are stuck in the TIME-WAIT state on Linux:

```
netstat -na | grep TIME
```

As long as one of these connections is hanging on, it prevents any new process from calling bind() on the same local port and address.

This failure of the bind() call can be prevented by setting the SO_REUSEADDR flag on the server socket before calling bind().

The following code demonstrates this:

```
int yes = 1;
if (setsockopt(my_socket, SOL_SOCKET, SO_REUSEADDR,
        (void*)&yes, sizeof(yes)) < 0) {
    fprintf(stderr, "setsockopt() failed. (%d)\n", GETSOCKETERRNO());
}
```

Once the SO_REUSEADDR flag is set, bind() succeeds even if a few TIME-WAIT connections are still hanging on to the same local port and address.

An example program, server_reuse.c, is included to demonstrate this technique.

I suggest that you always use SO_REUSEADDR for TCP servers because there are few downsides. The only real drawback is that using SO_REUSEADDR allows your program to bind to a specific interface even if another program has already bound to the wildcard address. Usually, this isn't a problem, but it is something to keep in mind.

You may sometimes see programs that attempt to fix this issue by killing sockets in the TIME-WAIT state. This can be accomplished by setting the socket linger option. This is dangerous! The TIME-WAIT state is essential to TCP's reliability, and interfering with it can lead to severe problems.

Why is this address-in-use only a problem for servers and not clients? Because the problem manifests itself when calling bind(). Client programs don't usually call bind(). If they do, this can also be a problem on the client-side.

While we are still on the topic of disconnected sockets, what happens when you try to send data to a peer that has already called close()? Let's consider that next.

Sending to a disconnected peer

There are three basic ways a TCP connection can fail. They are as follows:

- A network outage
- The peer application crashes
- The peer's system crashes

A network outage prevents data from reaching your peer. In this case, TCP tries to retransmit data. If connectivity is re-established, TCP simply picks back up where it left off. Otherwise, the connection eventually times out. This timeout can be on the order of 10 minutes.

The second way a TCP connection can fail is if the connected peer application crashes. In this case, the peer's operating system sends a FIN message. This case is indistinguishable from the peer calling close() on their end. If your application continues to send data after having received the FIN message, the peer's system will send an RST message to indicate an error.

Finally, a connection could fail because the peer's whole system has crashed. In this case, it won't be able to send a FIN message. This case looks the same as a network outage, and the TCP connection would eventually timeout. However, consider what happens if the crashed system reboots before the connection times out. In that case, the rebooted system will eventually receive a TCP message from the original connection. The rebooted system will not recognize the TCP connection and will send an RST message in response to indicate an error condition.

To reiterate, if you use send() on a socket that your peer thinks is closed, that peer will respond with an RST message. This state is easily detected by the return value of recv().

A more serious issue to consider is what happens when send() is called on a socket that has already received an RST message from its peer. On Unix-based systems, the default is to send a SIGPIPE signal to your program. If you don't handle this signal, the operating system will terminate your program.

It is therefore essential for TCP servers to either handle or disable the SIGPIPE signal. Failure to handle this case means that a rude client could kill your server.

Signals are complicated. If you're already using signals in your program, you may want to handle SIGPIPE. Otherwise, I recommend you just disable it by setting the SIGPIPE handler to SIG_IGN.

The following code disables SIGPIPE on Unix-based systems:

```
#if !defined(_WIN32)
#include <signal.h>
#endif

#if !defined(_WIN32)
signal(SIGPIPE, SIG_IGN);
#endif
```

As an alternative, you can use MSG_NOSIGNAL with send() as shown in the following code:

```
send(my_socket, buffer, sizeof(buffer), MSG_NOSIGNAL);
```

If the signal is ignored or MSG_NOSIGNAL is used, send() returns -1 and sets errno to EPIPE.

On Windows, attempting to call `send()` on a closed socket generally results in `WSAGetLastError()` returning `WSAECONNRESET`.

An example program, `server_crash.c`, is included in this chapter's code repository. This program accepts TCP connections on port `8080`. It then waits for the client to disconnect, and then attempts two sends to that disconnected client. This program is useful as a tool to explore the return values, error codes, and function behavior in different scenarios.

Socket's local address

When implementing servers, for both TCP and UDP, it is important to bind the listening socket to a local address and port. If the socket isn't bound, then clients can't know where to connect.

It is also possible to use `bind()` on the client side to associate a socket with a particular address and port. It is sometimes useful to use `bind()` in this manner on machines that have multiple network interfaces. The use of `bind()` can allow the selection of which network address to use for the outgoing connection.

Sometimes, `bind()` is used to set the local port for an outgoing connection. This is usually a bad idea for a few reasons. First, it very seldom serves any purpose. The port number presented to the connected server is likely to be different because of network address translation regardless. Binding to a local port also invites the error of selecting a port that is already in use. Usually, the operating system takes care of selecting a free port. This use of `bind()` also raises the issue with `TIME-WAIT`, which would prevent a new connection from being established after a closed one without a substantial delay.

We have used `bind()` in this book mostly for binding servers to a particular port number. It can also be used to associate servers to a particular address. If a server has multiple network interfaces, it may be the case that you only care to listen on connections at one address. In this case, `bind()` can easily be used to limit the connections to that address. It can also be used to limit connections to the local machine by binding sockets to `127.0.0.1`. This can be an important security measure for some applications.

We have employed the `select()` function for many purposes—timing out `connect()`, signaling when data is available, and preventing `send()` from blocking. However, `select()` is only suitable for monitoring a limited number of sockets. Let's look at this limitation, and how to circumvent it, next.

Multiplexing with a large number of sockets

We've used `select()` in this book to multiplex between open sockets.
The `select()` function is great because it is available on many platforms. However, if you have a large number of open sockets, you can quickly run into the limitations of `select()`.

There is a maximum number of sockets you can pass to `select()`. This number is available through the `FD_SETSIZE` macro.

This chapter's code repository includes a program, `setsize.c`, which prints the value of `FD_SETSIZE`.

The following screenshot shows this program being compiled and run on Windows 10:

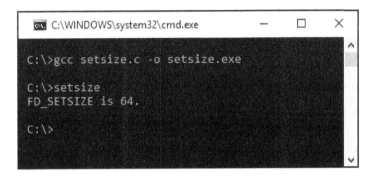

The preceding screenshot shows `FD_SETSIZE` is `64` on this system. Although Windows's default size for `FD_SETSIZE` is quite low, it is common to see higher values on other systems. The default value of `FD_SETSIZE` on Linux is `1024`.

On Windows, it is possible to increase `FD_SETSIZE` easily. You only need to define `FD_SETSIZE` yourself before including the `winsock2.h` header. For example, the following code increases `FD_SETSIZE` to `1024` on Windows:

```
#ifndef FD_SETSIZE
#define FD_SETSIZE 1024
#endif
#include <winsock2.h>
```

This works because Winsock uses `FD_SETSIZE` to build the `fd_set` type.

On Linux, this trick does not work. Linux defines `fd_set` as a bitmask, and it is not possible to increase its size without recompiling the kernel.

There are possible workarounds to effectively cheat `select()` into accepting socket descriptors larger than `1023` on Linux. One trick that usually works is to allocate an array of `fd_set` variables. Setting a socket is then done like this:

```
FD_SET(s % FD_SETSIZE, &set_array[s / FD_SETSIZE])
```

However, if you have to resort to a hack such as the preceding code, you may be better off avoiding `select()` and using a different multiplexing technique. The `poll()` function, for example, provides the functionality of `select()` without a limit on the number of file descriptors it can handle.

Summary

We covered a lot of ground in this chapter. First, we reviewed error-handling methods, and then we implemented a function to obtain text descriptions for error codes.

We then jumped right into the hard details of TCP sockets. We saw how TCP sockets hide much complexity, and how it is sometimes necessary to understand that hidden state to get good application performance. We saw a method for an early timeout on a TCP `connect()` call, and we looked at how to terminate a connection with an orderly release.

We then took a closer look at the `bind()` function and how its usefulness differs between servers and clients. Finally, we discussed how the `select()` function limits the total number of sockets your program can handle, and how to work around it.

So far, this book has been focused mainly on network code as it would pertain to personal computers and servers. In the next chapter, Chapter 14, *Web Programming for the Internet of Things*, we move our focus to the extending of internet access to everyday objects—that is, the **Internet of Things**.

Questions

Try these questions to test your knowledge acquired from this chapter:

1. Is it ever acceptable to just terminate a program if a network error is detected?
2. Which system functions are used to convert error codes into text descriptions?
3. How long does it take for a call to `connect()` to complete on a TCP socket?
4. What happens if you call `send()` on a disconnected TCP socket?
5. How can you ensure that the next call to `send()` won't block?
6. What happens if both peers to a TCP connection try to send a large amount of data simultaneously?
7. Can you improve application performance by disabling the Nagle algorithm?
8. How many connections can `select()` handle?

The answers to these questions are found in `Appendix A`, *Answers to Questions*.

14
Web Programming for the Internet of Things

In this chapter, we turn our attention to the **Internet of Things** (**IoT**). The IoT is an exciting new trend where internet connectivity is added to everyday physical objects. When combined with embedded electronics and sensors, internet access allows physical objects to interact with one another and be controlled and monitored from anywhere in the world.

The following topics are covered in this chapter:

- Defining the IoT
- Connectivity types
- Bandwidth considerations
- Controller types
- Ethics of the IoT
- Security for the IoT

Technical requirements

This chapter contains no C code. Instead, it focuses on the theory, techniques, and practices used in the IoT domain.

What is the IoT?

We now live in a world where almost anything that can be connected to the internet, has been. These devices make up the IoT, and they are in every part of our lives.

In the kitchen at home, devices with internet connectivity include fridges, microwave ovens, conventional ovens, food scales, dishwashers, coffee-makers, and even juicers.

Elsewhere in the home, we have smart TVs, gaming consoles, thermostats, furnaces, washing machines, light switches, light bulbs, yoga mats, alarm clocks, cameras, doorbells, bathroom scales, baby monitors, sound systems, and speakers—all connected to the internet. Of course, your whole house may be supplied with electricity, water, and gas from networked smart meters. When you leave home, do you go through an IoT garage door to get in your always-connected car with built-in Wi-Fi? Many do.

In the industrial sector, IoT devices are central to **Industry 4.0**, commonly considered the fourth industrial revolution. Using internet connectivity to interconnect every device in the manufacturing supply chain has allowed for unprecedented optimizations in efficiency.

The IoT is even affecting the way we grow food. In agriculture, IoT devices allow farmers to monitor relevant weather conditions, such as temperature, humidity, sunlight, rainfall, and wind speed. This information informs decisions improving yield and quality.

Of course, these are just a few examples of IoT applications. The concept is also applied to concerns in environment, healthcare, infrastructure, transportation, and more.

When adding internet connectivity to a device, you have several options. There are certainly pros and cons to each connection option. We review them next.

Connectivity options

Once you've figured out what your device does, and that it benefits from an internet connection, you still have to decide how to connect it. There are many options with various trade-offs.

Wi-Fi

Perhaps Wi-Fi needs no introduction. Almost every modern home with internet access has Wi-Fi available. This is in large part because residential **Internet Service Providers** (**ISPs**) typically use modems with built-in Wi-Fi. If your IoT device is deployed indoors, there is a good chance that a Wi-Fi network is available. This is particularly true for residential and commercial devices.

Wi-Fi is the most popular **Wireless Local Area Network** (**WLAN**) technology and is based around the IEEE 802.11 standards. It typically operates on either a 2.4 GHZ or a 5 GHZ radio frequency.

Two main operation modes are provided by Wi-Fi. In **ad hoc mode**, devices communicate directly to one another. However, **infrastructure mode** is much more common. In infrastructure mode, devices on a WLAN all connect to a **Wireless Access Point** (**WAP**). The WAP then generally provides internet access for the whole WLAN.

Wi-Fi provides good bandwidth, low latency, and basic security features, but it has higher power needs than other short-range wireless standards. Although newer versions of Wi-Fi can achieve speeds of 1 Gbps, these are not yet commonly seen in practice. In any case, for an IoT device application, we are typically concerned about internet access. The Wi-Fi performance of the IoT device's connection to the local router is not usually the bottleneck, but rather, the available internet service to the router is.

One downside to using Wi-Fi connectivity in your IoT device is in the setup difficulty. Most Wi-Fi networks are secured with a password, and therefore your device needs to somehow obtain the network name and password before being able to connect. If your device has a built-in screen and keyboard (or touchscreen), then entering the network password is trivial.

However, for IoT devices without large screens, getting the network password configured can sometimes be complicated. One solution is to have your end user put the configuration information on an SD card or USB storage device. Your device then reads the password from this storage and can connect to the network. The problem with this solution is that it is tedious, and many users may not be technically capable of completing this setup.

An alternative setup is to first connect the device to a computer or smartphone (perhaps through Bluetooth) for the initial configuration. This has the downside of requiring the user to obtain such a setup device (not everyone has a smartphone), as well as requiring extra work in developing the setup application.

A third setup method is to have the IoT device initially provide its own Wi-Fi network. In other words, the device acts as a temporary access point. The user then utilizes a laptop or smartphone to connect directly to the device. The user opens a browser, and the device serves a web page on which the user can configure the device. Once the parameters are entered, the IoT device connects to the local Wi-Fi hotspot and is now configured for internet access.

The following diagram illustrates this setup method:

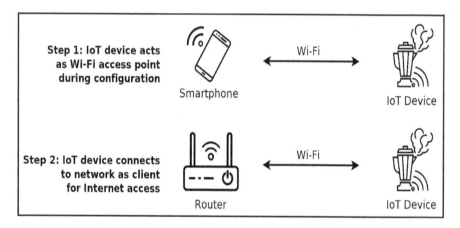

The advantage of this setup method is that no special software needs to be developed for the setup. Any device that can connect to a Wi-Fi network and has a standard web browser will work.

Keep in mind that Wi-Fi networks change occasionally. It is good security practice to change Wi-Fi passwords from time to time, and each time this happens, the IoT device will need to be reconfigured.

Although Wi-Fi is convenient, if a wired connection is available, it's often easier and better.

Ethernet

A wired Ethernet connection is ideal for any IoT device, if it's available. Ethernet provides easy setup and extreme reliability. The newest standards allow for a bandwidth of 400 Gbps, and its low latency is unrivaled. It is even possible to provide power over Ethernet in some cases, further simplifying deployment.

The downside to Ethernet connections is, of course, that a physical cable is required. Many offices already have Ethernet run to each room, and Ethernet-connected IoT devices work well there. In residential and industrial settings, however, running new wires can be tedious. Wireless connections are, therefore, often easier in those settings.

When available, Ethernet setup is easy. Oftentimes, a device only needs to be plugged in and it's ready to work.

Keep in mind that, with Ethernet, as with Wi-Fi, you're relying on internet access being available at your IoT device installation site. This internet access is generally much slower than the Ethernet connection you will have on the local network. As this internet access is provided by your customer, its quality can vary greatly.

What if Wi-Fi and Ethernet aren't available? What if your device needs to be mobile? In those cases, cellular internet access may be your best option.

Cellular

Cellular (mobile networks) can be a great connection option for IoT devices. It is particularly useful for two reasons. First, cellular service is available nearly everywhere. This makes cellular ideal for mobile IoT devices. Second, all network setup can be completed at the time of manufacture. This makes deployment easy, hassle-free, and can lower ongoing support costs when compared to other options.

The drawback to cellular connectivity is higher cost. First, the IoT device must contain a cellular modem. Cellular modems are costlier than Wi-Fi modems and Ethernet transceivers. In addition, certain testing and certification is required before a device is allowed on operator networks.

Cellular devices must also pay for access. With Wi-Fi or Ethernet, an IoT device is assumed to be able to leech off local internet access. With cellular, someone must pay the bill for each specific device. Usually, it makes sense for the IoT device manufacturer to provision and maintain service for each device. For devices requiring low bandwidth, an LTE cellular connection can be maintained for well under $1.00 (USD) per device per month. Any device that requires significant bandwidth is much more costly to service.

An additional benefit of cellular is that many mobile network providers can route your traffic through a **Virtual Private Network** (**VPN**). This is done by the network operator, and you won't be billed for the additional bandwidth of the VPN overhead. A VPN isolates your IoT devices from the internet and allows communication only with your trusted servers. It is convenient because many of your security concerns are taken care of by the cellular provider.

We've covered three options for direct internet access, but many IoT devices don't need to access the internet directly. Some work in conjunction with a computer or smartphone for access. One wireless technology commonly used with this method is Bluetooth.

Bluetooth

Bluetooth is a popular short-range wireless technology standard for use in **Wireless Personal Area Networks** (**WPAN**). Like Wi-Fi, Bluetooth operates on the 2.4 GHz radio band.

Bluetooth can be used for direct internet access, but this is uncommon in practice. Bluetooth is usually used to connect an IoT device to another device that has internet access. For example, many smartwatches use Bluetooth to pair with smartphones. The smartphone then has internet access through either a Wi-Fi or cellular connection.

The following diagram shows a Bluetooth **smart watch** connection:

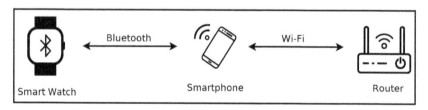

This works very well for the smart watch, as it is usually expected to be in close proximity to a **smartphone**.

Of course, you may object philosophically to the smart watch being considered an IoT device. In this case, it's more of a sensor for a smartphone app than a standalone IoT device. Nonetheless, it fits into the ecosystem of the IoT. This type of piggy-back connectivity can be a good option and should not be overlooked.

As a similar technology to Bluetooth, **IEEE 802.15.4**-based wireless networks are commonly deployed to allow a whole network of IoT devices to intercommunicate.

IEEE 802.15.4 WPANs

IEEE 802.15.4 is a technical standard for **low-rate wireless personal area networks** (**LR-WPAN**). The standard was first defined in 2003, and it provides the foundation that other protocols are built on. For example, the **Zigbee**, **6LoWPAN**, **Thread**, **ISA100.1**, **WirelessHART**, **WiSUN**, and **MiFi** protocols are all based on the IEEE 802.15.4 standard, just to name a few.

IEEE 802.15.4-based protocols are useful for low-power, limited-range applications. Typical ranges extend to about 10 meters with a transfer rate of 250 Kbps. Because of the low power requirements, battery-powered devices are often able to operate for multiple years on a single charge.

Some protocols built on IEEE 802.15.4, such as 6LoWPAN, allow devices to communicate using IP directly, similar to how Wi-Fi is typically used. Other protocols are commonly used only to send data to a central hub device. This hub is then responsible for processing IoT device data and communicating over the internet, if needed.

As a concrete example, home automation networks are often built on the IEEE 802.15.4 standard. In many cases, these networks are seldom used to provide direct internet access to each IoT device. Instead, a hub is commonly used to allow a network of devices to communicate. The hub then coordinates the data and forwards it on to the internet.

A **mesh network topology** is common in IEEE 802.15.4-based setups. This allows devices remote from the hub to communicate through closer devices.

The following diagram illustrates a mesh network for home automation:

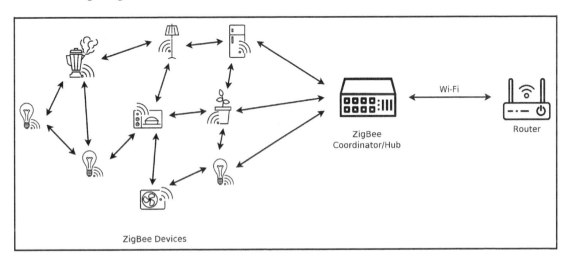

In the preceding diagram, we see an IEEE 802.15.4-based network for a home automation setup. Various IoT devices, such as lights and kitchen gadgets, communicate with a **Coordinator/Hub**. This hub controls and coordinates the connected devices. The hub itself uses Wi-Fi to connect to the local network, and this internet connection allows for remote control and monitoring of IoT devices.

Now that we've covered a few of the most popular connection options, let's turn our attention to IoT hardware choices.

Hardware choices

When designing an IoT device, there are many choices for hardware. In many cases, these choices ultimately must be made based on the functionality of the device, and not on their connectivity options. Remember, IoT devices need to have a useful purpose besides just connecting to the internet.

Let's look at three options for hardware—single-board computers, microcontrollers, and **Field-Programmable Gate Arrays** (**FPGAs**).

Single-board computers

Single-Board Computers (**SBCs**) are complete computers built on a single circuit board. They include all the usual pieces of a functional computer—the **Central Processing Unit** (**CPU**), **Random Access Memory** (**RAM**), non-volatile storage, and input/output ports.

SBCs are capable of running full-featured operating systems. Linux is commonly used.

The **Raspberry Pi Zero W**, for example, is an SBC measuring just 2.6 by 1.2 inches. It has 512 MB of RAM built-in and uses a micro SD card for storage. Wi-Fi and Bluetooth connectivity is also included. The Raspberry Pi Zero W currently retails for about $10 (USD).

The following photo shows a Raspberry Pi Zero W:

Because SBCs run normal desktop operating systems, such as Linux, programming them is easy. In fact, all of the network programs we've developed throughout this book are applicable.

In addition to standard computer connectors for video, audio, and USB, SBCs usually have some **General-Purpose Input/Output** (**GPIO**) pins to work with. This gives them some basic microcontroller-like functionality. These GPIO pins are capable of reading a digital input signal and providing a digital output signal. On the Raspberry Pi, some pins are also able to provide **Pulse Width Modulation** (**PWM**) outputs, as well as various board-level serial protocols—**Serial Peripheral Interface** (**SPI**), **Inter-Integrated Circuit** (**I2C**), and **Asynchronous Serial**.

Although SBCs are very easy to get started with, they do have a few drawbacks. They are relatively expensive, they require a lot of power, and their use of general-purpose operating systems isn't always ideal for embedded systems. Microcontrollers, by comparison, are much cheaper and optimized to handle real-time performance constraints.

Microcontrollers

A **microcontroller** is a small computer contained on a single integrated circuit. Microcontrollers contain all the parts needed for a functional computer, including a CPU, RAM, and non-volatile storage.

While SBCs usually have processor speeds in the hundreds or thousands of MHz, it is common for microcontrollers to be clocked at less than 100 MHz. It's not uncommon to see microcontrollers clocked at even 1 MHz or less. Microcontrollers often have RAM varying from between a few dozen bytes to a few megabytes. These humble specifications also mean that microcontrollers can get by on very little power. It's not uncommon to see microcontrollers needing only a few nano-amps when in sleep mode.

Microcontrollers usually come with lots of GPIO pins and useful peripherals, such as timers, real-time clocks, communication interfaces, **Pulse Width Modulation** (**PWM**) outputs, **Analog-to-Digital Converter** (**ADC**) comparators, **Digital-to-Analog Converter** (**DAC**), **Watch-Dog Timers** (**WDT**), serial transceivers, and more.

These microcontroller peripherals are often vital to the functioning of an IoT device core functionality. For example, a PWM generator peripheral can be used for precise motor control, while an ADC may be needed to read a sensor. Remember, the internet part of an IoT device is largely secondary; the device must perform some useful function first. What use is an internet-connected thermostat if it can't sense the temperature?

Microcontrollers come in various architectures and bus widths. 8-bit, 16-bit, and 32-bit controllers are all common. Although for IoT devices, 8-bit controllers are probably under-powered.

Modern desktop computers almost always use the Von Neumann architecture, but many microcontrollers use the Harvard architecture.

In the Von Neumann architecture, which dominates the desktop processor market, memory stores both data and instructions. This necessitates that processor instructions are the same width as the data they are processing. Harvard architecture machines, by contrast, use completely separate memories to store instructions and data. It is common for Harvard architecture microcontrollers to use different word sizes for instructions compared to data. For example, a 16-bit Harvard architecture microcontroller may process data in 16-bit words, but its instructions could be encoded as 24-bit words. While data is stored in RAM, instructions are read directly from non-volatile memory, such as **FLASH** or **EEPROM**.

It is also possible to find microcontrollers with built-in networking functionality, including support for Wi-Fi, Ethernet, Bluetooth, IEEE 802.15.4, and others.

As an example, the ESP8266 32-bit microcontroller, produced by Espressif Systems, comes with Wi-Fi transceiver and TCP/IP stack.

The following diagram shows an ESP8266 microcontroller on a breakout board:

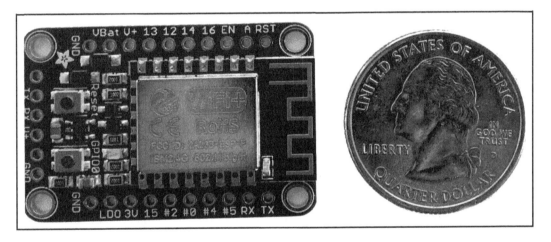

Microcontrollers that don't have built-in transceivers can be connected to external modems with those capabilities.

Microcontrollers are often programmed directly, without an operating system. Some programmers prefer operating systems, and many different microcontroller operating systems are available. These operating systems often make it easier for a microcontroller to perform various tasks while simplifying program layout and retaining real-time capabilities. It should be noted that these microcontroller operating systems generally lack memory segmentation and other safety features taken for granted in desktop operating systems.

Microcontrollers are great when you don't need much processing power, but do require real-time processing guarantees. What about when you need both processing power and real-time guarantees? See FPGAs.

FPGAs

Field-Programmable Gate Arrays (**FPGAs**) are integrated circuits containing arrays of programmable logic blocks. These logic blocks can be programmed and combined to implement complex functions. Essentially, any logic that an **Application-Specific Integrated Circuit** (**ASIC**) can perform can also be done on an FPGA.

FPGAs are useful for highly demanding processing tasks requiring real-time guarantees, especially tasks that benefit from highly-parallel processing. For example, a video-processing algorithm that is too demanding for a single-board computer, much less a microcontroller, may be trivial for an FPGA.

FPGAs are programmed using **Hardware Description Languages** (**HDL**). The two most common languages used are **Verilog** and **VHSIC Hardware Description Language** (**VHDL**). These languages are much different than C. While C is an imperative language that gives the steps to take to complete a calculation, HDLs are declarative languages, giving descriptions of how logic gates should be connected.

In practice, it is common to implement a soft microprocessor core in an FPGA. This soft processor can then be programmed like any other microprocessor and is used for sundry tasks that don't require much processing power. For example, an FPGA that processes video may implement a microprocessor that is responsible for the setup and configuration functions of the system. These soft processors are commonly programmed in C.

Regardless of which type of hardware your IoT device uses, if it doesn't have suitable built-in communication, you will need an external transceiver or modem.

External transceivers and modems

External transceivers and modems are available for Wi-Fi, cellular, Ethernet, Bluetooth, and any other protocol you can think of.

The following photo shows an industrial IoT device that uses a cellular modem:

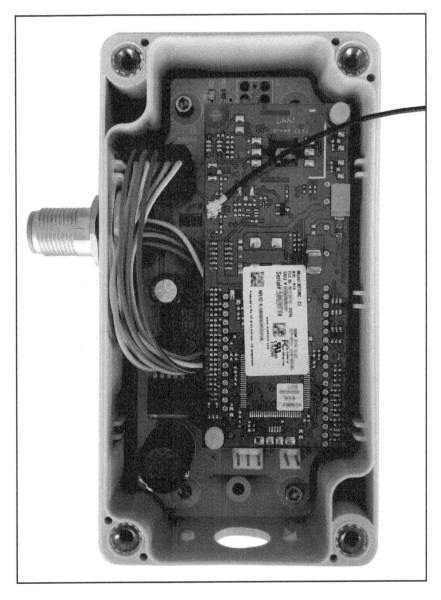

In the preceding diagram, the lower circuit board implements the functionality of the device (monitoring cryogenic liquids, in this case). On this board, the power supply system, sensor electronics, and a low-cost, 16-bit microcontroller are found. A cellular modem is located on the top circuit board, which plugs into a socket on the lower board.

External modems often communicate with their host (microcontroller) through asynchronous serial. The protocols used vary considerably among systems.

One common protocol family for communicating with modems is the **Hayes command set** also called **AT commands**. These commands consist of short text-based messages. AT commands were more or less standardized for placing phone calls, but have been extended to managing internet connections. Unfortunately, almost every modem manufacturer does the extension differently.

If you're using an external modem, you need to consult that modem's documentation for the exact protocol used to communicate with it. Socket-style drivers are often not available, but many of the core networking concepts explored in this book still apply.

With connectivity out of the way, let's turn our attention to **IoT device protocols**.

IoT protocols

Most IoT devices work by sending their data to a few central servers—the cloud. These servers process and store the IoT device data and allow remote access and configuration.

For example, take a smart thermostat. It continuously sends temperature data to a central server. This server stores the data. If a user wants to view the data, they connect their personal computer or smartphone to that central server. They don't connect directly to the IoT device itself. When they want to change their thermostat settings, they send this change to the central server, which then relays it to the IoT thermostat.

The following diagram illustrates this concept:

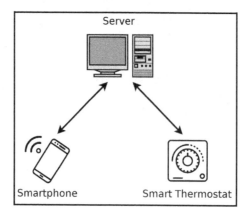

In the preceding diagram, note that all communication goes through the server. There is never direct communication from the smartphone to the IoT device.

Using a central server has several advantages. It allows the IoT device to do much less processing. The server can further process that data and format it into graphs and such. The server can also send emails or other alerts, and the server is able to store much more data than the IoT device is capable of. Furthermore, the server gives both the IoT device and the user's device (a laptop or smartphone) a place to connect to. Without a central server, getting the user's device to connect to the IoT device would be its own challenge.

The actual protocols used by IoT devices vary. Many devices use HTTPS. In these cases, the device usually requests an HTTPS page while passing collected data. The web server will then collect the data and return any needed information to the IoT device.

There are a few protocol standards aimed specifically for IoT devices, such as **Constrained Application Protocol** (**CoAP**) and **Message Queuing Telemetry Transport** (**MQTT**).

However, custom designed TCP and UDP protocols are also common. These custom protocols can achieve a much higher level of efficiency where bandwidth is a concern.

Perhaps we will see more protocol standardization in the coming years. This could have an enormous advantage in allowing devices to better interoperate. As it is now, devices rarely interoperate with devices from different manufacturers.

Another important consideration when deploying an IoT device is how to update the code after the device is shipped.

Firmware updates

IoT devices are connected to the internet, and this gives us an advantage for product updates. When new functionality or bug fixes are developed, we can use the internet to push updates to our IoT devices.

This process is easy when using an SBC. You just push the update as you would for any software.

If your product uses a microcontroller or FPGA, then things get a bit more complicated. Your device will need to download a firmware image, and then transfer that into non-volatile memory as required.

If the device's power is interrupted during the firmware update phase, the device could be left in an unusable state. This can be prevented through careful design. If the device has enough memory to store the firmware twice, it is possible to download the entire firmware update without overwriting the original. In this case, the device can detect the failed update (by using a checksum, watchdog timer, or someother mechanism) and revert to the previous state.

In any case, there are a few practices that should be followed by any IoT device doing remote updates.

First, it is essential for security that any firmware update is authenticated. If this step is missed, then an attacker could send a fake update to your device. Once your device is running the attacker's code, nothing else you do for security matters. We will discuss security in more detail later in this chapter.

Despite the focus on authentication, you should consider allowing your device's owner to install unauthenticated firmware. It is a basic user freedom to run the code they want to run on their own devices, and some consider it unethical to attempt to prevent this. In any case, it is not technically possible to retain control of a device that's not physically under your control. In other words, if you attempt to prevent a user from installing their own firmware through **Digital Rights Management** (**DRM**), you should expect your efforts to be eventually defeated.

It is often considered good practice to allow the device to return to its original factory configuration. This usually means storing the factory firmware in a separate non-volatile memory, such as a low-cost flash chip. This is important in the rare case that a firmware update fails. If a firmware update fails, the device may be unable to boot. If a redundant firmware copy was stored, then the device can be rolled back to its original factory state with some sort of physical reset mechanism. From this state, the update can be retried.

Now that we've covered some technical considerations of IoT design, let's turn our attention to a few ethical concerns.

Ethics of IoT

Much criticism has been leveled at the IoT concept regarding privacy, security, and other ethical concerns. This isn't surprising, given these devices' pervasive presence in our lives.

Most IoT devices work in conjunction with a central server. This server gets essentially all of the collected data from the IoT devices it services. This raises questions about data ownership and privacy.

Privacy and data collection

Many IoT devices collect lots of data as part of their operation. For example, a smart thermostat collects temperature data about its environment. This is required for it to function. This data may seem harmless at first glance, but once you consider how people use smart thermostats, you realize that the data is more important than it first appears.

In the smart thermostat example, people will set different temperatures based on when they expect to be awake, asleep, or away. From this data, you can infer approximately at what time someone leaves for work in the morning and what time they get back. When a person leaves their house for a week-long holiday, it will be clearly reflected in the data.

If even a seemingly innocuous smart thermostat gives great insight into a person's behavior, imagine how much more data a smartphone or home assistant collects!

If we accept that this data collection is inherent to the device's functionality, then its collection seems justified. Consider, then, if the IoT company has an obligation to the user to keep their data secure or confidential. Then who owns this data? Is it owned by the company collecting it, or does the customer who provides that data own it? The old understanding was that the data was solely the property of the company collecting it, but there is a recent push to reclassify that. Legal opinions vary, and this is still mostly uncharted territory.

In any case, if you're collecting data, even seemingly innocuous data, please treat it with respect. Please also allow your users the freedom to see and download copies of that data you are storing on them.

End-of-life planning

Another consideration when designing an IoT device is what happens when that device is no longer being manufactured? Many IoT companies will simply stop supporting their old devices after a while, and these devices then become useless. This is because they often require connecting to a central server to function, and once this central server goes offline, they literally cannot function.

There is an alternative. When designing an IoT device, it is possible to make it function in a reduced capacity even without internet connectivity. For example, it is possible to design a smart thermostat to continue working as a manual thermostat even when it cannot establish a connection to its central server.

When designing your IoT device, please take into consideration if you can keep it useful even after your company discontinues support for it. In some cases, this isn't possible, but in many cases, it appears that companies purposely limit their devices only to spite their customers. Don't be like those companies.

Security

In recent years, a lot of IoT devices have been deployed with little thought given to security. As companies rush to push devices to market quickly, security too often doesn't even enter the equation. The situation has gotten so bad that there is now a common saying: *"The S in IoT stands for* security.*"*

The security problem may be compounded by the fact that insecure IoT devices don't often expose the device developers to liability. For example, if you develop an insecure server, then your company is open to attack. However, if you develop an insecure IoT device, your company is fine; it's just your customers who are open to attack.

If you're developing an IoT device, I implore you—do the right thing and think about security early and often.

According to the **Open Web Application Security Project (OWASP)**, the top ten IoT security problems are as follows:

1. Weak, guessable, or hardcoded passwords
2. Insecure network services
3. Insecure ecosystem interfaces
4. Lack of secure update mechanism
5. Use of insecure or outdated components
6. Insufficient privacy protection
7. Insecure data transfer and storage
8. Lack of device management
9. Insecure default settings
10. Lack of physical hardening

You can find more information from the OWASP Foundation at https://owasp.org/.

Many of the preceding problems are easily addressable simply by caring. For example, many devices are still shipped with hardcoded passwords or backdoors. There isn't much to say about this problem from a technical perspective—just don't ship your products with hardcoded passwords or backdoors.

Some advocate that all IoT devices should use HTTPS to provide a basic level of security—the so-called **Secure Hypertext Internet of Things (SHIoT)**. See `Chapter 9`, *Loading Secure Web Pages with HTTPS and OpenSSL*, for information about HTTPS from the client's perspective.

For embedded HTTPS, you may find the OpenSSL library too heavyweight. Consider using a TLS implementation designed for embedded use. These embedded-first libraries are also able to take advantage of encryption hardware built in to common microcontrollers. **wolfSSL** is one such open-source library. In addition to TLS for TCP, wolfSSL also supports DTLS for securing UDP-based protocols.

Using TLS can go a long way to secure an IoT device, but it is not a panacea. In particular, certificates must be carefully managed. Don't make the mistake of assuming that encryption is good enough. Encryption only protects against passive attacks; both encryption and authentication are needed to protect against active attacks.

Sometimes, minimizing bandwidth is a priority. This is particularly true for cellular connections. In many use-cases, an IoT device with limited bandwidth will find TLS and DTLS too heavyweight. Establishing a new TLS or DTLS connection can take several kilobytes on average.

If your cellular provider is able to tunnel your traffic over a secure VPN, without metering the VPN overhead, you may be able to forgo implementing secure communication at the application level. In this case, your device can use TCP or UDP directly, and the VPN connection ensures that this traffic will never be visible over the internet. This level of security isn't appropriate for all use-cases, as your traffic is still visible to your network provider.

In the case that you must secure a protocol yourself, you may take advantage of the fact that you have control over the manufacturing of your IoT device. It is possible to assign each device a unique encryption key at the time of manufacture. This technique is called **Pre-Shared Key (PSK)**. With PSK, devices can secure their communication using symmetric ciphers, avoiding much of the overhead required by TLS, DTLS, and other heavyweight solutions. If doing symmetric encryption, keep in mind that encryption is useless against an active attacker unless you also implement authentication. This is done using a **Message Authentication Code (MAC)**.

If you do use **PSK**, be sure to use a unique key for each individual device. Reusing the same key is not appropriate.

In any case, it is advisable to run your security scheme by a security professional before and during implementation. Even a small, nearly insignificant mistake with encryption or authentication can compromise an entire system.

Summary

In this chapter, we looked at the high-level design of IoT devices. We looked at different types of hardware that IoT devices can be built on—SBCs, microcontrollers, and FPGAs. We considered the various ways these devices can connect to the internet, including wired and wireless connections, and we weighed some trade-offs associated with each.

We also took into consideration some of the protocol choices that IoT devices can use and reiterated the importance of security.

Questions

Try these questions to test your knowledge from this chapter:

1. What are the drawbacks of using Wi-Fi connectivity?
2. What are the drawbacks of using Ethernet connectivity?
3. What are the drawbacks of using cellular connectivity?
4. What are some advantages of using a Single-Board Computer with embedded Linux? What are the drawbacks?
5. What are some advantages of using a microcontroller in your IoT device?
6. Is the use of HTTPS always appropriate in IoT device communication?

The answers to these questions can be found in `Appendix A`, *Answers to Questions*.

Answers to Questions

Chapter 1, Introducing Networks and Protocols

1. **What are the key differences between IPv4 and IPv6?**

   IPv4 only supports 4 billion unique addresses, and because they were allocated inefficiently, we are now running out. IPv6 supports 3.4×10^{38} possible addresses. IPv6 provides many other improvements, but this is the one that affects our network programming directly.

2. **Are the IP addresses given by the** `ipconfig` **and** `ifconfig` **commands the same IP addresses that a remote web server sees if you connect to it?**

   Sometimes, these addresses will match, but not always. If you're on a private IPv4 network, then your router likely performs network address translation. The remote web server then sees the translated address.
   If you have a publicly routable IPv4 or IPv6 address, then the address seen by the remote web server will match those reported by `ipconfig` and `ifconfig`.

3. **What is the IPv4 loopback address?**

   The IPv4 loopback address is `127.0.0.1`, and it allows networked programs to communicate with each other while executing on the same machine.

4. **What is the IPv6 loopback address?**

   The IPv6 loopback address is `::1`. It works in the same way as the IPv4 loopback address.

5. **How are domain names (for example,** `example.com`**) resolved into IP addresses?**

   DNS is used to resolve domain names into IP addresses. This protocol is covered in detail in `Chapter 5`, *Hostname Resolution and DNS*.

6. **How can you find your public IP address?**

The easiest way is to visit a website that reports it for you.

7. **How does an operating system know which application is responsible for an incoming packet?**

Each IP packet has a local address, remote address, local port number, remote port number, and protocol type. These five attributes are memorized by the operating system to determine which application should handle any given incoming packet.

Chapter 2, Getting to Grips with Socket APIs

1. **What is a socket?**

A socket is an abstraction that represents one endpoint of a communication link between systems.

2. **What is a connectionless protocol? What is a connection-oriented protocol?**

A connection-oriented protocol sends data packets in the context of a larger stream of data. A connectionless protocol sends each packet of data independently of any before or after it.

3. **Is UDP a connectionless or connection-oriented protocol?**

UDP is considered a connectionless protocol. Each message is sent independently of any before or after it.

4. **Is TCP a connectionless or connection-oriented protocol?**

TCP is considered a connection-oriented protocol. Data is sent and received in order as a stream.

5. **What types of applications generally benefit from using the UDP protocol?**

UDP applications benefit from better real-time performance while sacrificing reliability. They are also able to take advantage of IP multicasting.

6. **What types of applications generally benefit from using the TCP protocol?**

Applications that need a reliable stream of data transfer benefit from the TCP protocol.

7. **Does TCP guarantee that data will be transmitted successfully?**

TCP makes some guarantees about reliability, but nothing can truly guarantee that data is transmitted successfully. For example, if someone unplugs your modem, no protocol can overcome that.

8. **What are some of the main differences between Berkeley sockets and Winsock sockets?**

The header files are different. Sockets themselves are represented as signed versus unsigned `int`s. When `socket()` or `accept()` calls fail, the return values are different. Berkeley sockets are also standard file descriptions. This isn't always true with Winsock. Error codes are different and retrieved in a different way. There are additional differences, but these are the main ones that affect our programs.

9. **What does the `bind()` function do?**

The `bind()` function associates a socket with a particular local network address and port number. Its usage is almost always required for the server, and it's usually not required for the client.

10. **What does the `accept()` function do?**

The `accept()` function will block until a new TCP client has connected. It then returns the socket for this new connection.

11. **In a TCP connection, does the client or the server send application data first?**

Either the client or the server can send data first. They can even send data simultaneously. In practice, many client-server protocols (such as HTTP) work by having the client send a request first and then having the server send a response.

Chapter 3, An In-Depth Overview of TCP Connections

1. **How can we tell if the next call to** `recv()` **will block?**

   We use the `select()` function to indicate which sockets are ready to be read from without blocking.

2. **How can you ensure that** `select()` **doesn't block for longer than a specified time?**

   You can pass `select()` a timeout parameter.

3. **When we used our** `tcp_client` **program to connect to a web server, why did we need to send a blank line before the web server responded?**

   HTTP, the web server's protocol, expects a blank line to indicate the end of the request. Without this blank line, it wouldn't know if the client was going to keep sending additional request headers.

4. **Does** `send()` **ever block?**

   Yes. You can use `select()` to determine when a socket is ready to be written to without blocking. Alternatively, sockets can be put into non-blocking mode. See `Chapter 13`, *Socket Programming Tips and Pitfalls*, for more information.

5. **How can we tell if a socket has been disconnected by our peer?**

   The return value of `recv()` can indicate if a socket has been disconnected.

6. **Is data received by** `recv()` **always the same size as data sent with** `send()`**?**

   No. TCP is a stream protocol. There is no way to tell if the data returned from one `recv()` call was sent with one or many calls to `send()`.

7. Consider this code:

```
recv(socket_peer, buffer, 4096, 0);
printf(buffer);
```

What is wrong with it?
Also see what is wrong with this code:

```
recv(socket_peer, buffer, 4096, 0);
printf("%s", buffer);
```

The data returned by recv() is not null terminated! Both of the preceding code excerpts will likely cause printf() to read past the end of the data returned by recv(). Additionally, in the first code example the data received could contain format specifiers (for example %d), which would cause additional memory access violations.

Chapter 4, Establishing UDP Connections

1. **How do** sendto() **and** recvfrom() **differ from** send() **and** recv()**?**

The send() and recv() functions are useful after calling connect(). They only work with the one remote address that was passed to connect(). The sendto() and recvfrom() functions can be used with multiple remote addresses.

2. **Can** send() **and** recv() **be used on UDP sockets?**

Yes. The connect() function should be called first in that case. However, the sendto() and recvfrom() functions are often more useful for UDP sockets.

3. **What does** connect() **do in the case of a UDP socket?**

The connect() function associates the socket with a remote address.

4. **What makes multiplexing with UDP easier than with TCP?**

One UDP socket can talk to multiple remote peers. For TCP, one socket is needed for each peer.

5. **What are the downsides to UDP when compared to TCP?**

UDP does not attempt to fix many of the errors that TCP does. For example, TCP ensures that data arrives in the same order it was sent, TCP tries to avoid causing network congestion, and TCP attempts to resend lost packets. UDP does none of this.

6. **Can the same program use UDP and TCP?**

Yes. It just needs to create sockets for both.

Chapter 5, Hostname Resolution and DNS

1. **Which function fills in an address needed for socket programming in a portable and protocol-independent way?**

   `getaddrinfo()` is the function to use for this.

2. **Which socket programming function can be used to convert an IP address back into a name?**

   `getnameinfo()` can be used to convert addresses back to names.

3. **A DNS query converts a name to an address, and a reverse DNS query converts an address back into a name. If you run a DNS query on a name, and then a reverse DNS query on the resulting address, do you always get back the name you started with?**

   Sometimes, you will get the same name back but not always. This is because the forward and reverse lookups use independent records. It's also possible to have many names point to one address, but that one address can only have one record that points back to a single name.

4. **What are the DNS record types used to return IPv4 and IPv6 addresses for a name?**

   The A record type returns an IPv4 address, and the AAAA record type returns an IPv6 address.

5. **Which DNS record type stores special information about email servers?**

The MX record type is used to return email server information.

6. **Does** getaddrinfo() **always return immediately? Or can it block?**

If getaddrinfo() is doing name lookups, it will often block. In the worst-case scenario, many UDP messages would need to be sent to various DNS servers, so this can be a noticeable delay. This is one reason why DNS caching is important.

If you are simply using getaddrinfo() to convert from a text IP address, then it shouldn't block.

7. **What happens when a DNS response is too large to fit into a single UDP packet?**

The DNS response will have the TC bit set in its header. This indicates that the message was truncated. The query should be resent using TCP.

Chapter 6, Building a Simple Web Client

1. **Does HTTP use TCP or UDP?**

HTTP runs over TCP port 80.

2. **What types of resources can be sent over HTTP?**

HTTP can be used to transfer essentially any computer file. It's commonly used for web pages (HTML) and the associated files (such as styles, scripts, fonts, and images).

3. **What are the common HTTP request types?**

GET, POST, and HEAD are the most common HTTP request types.

4. **What HTTP request type is typically used to send data from the server to the client?**

GET is the usual request type for a client to request a resource from the server.

5. **What HTTP request type is typically used to send data from the client to the server?**

POST is used when the client needs to send data to the server.

6. What are the two common methods used to determine an HTTP response body length?

The HTTP body length is commonly determined by the Content-Length header or by using Transfer-Encoding: chunked.

7. **How is the HTTP request body formatted for a** POST**-type HTTP request?**

This is determined by the application. The client should set the Content-Type header to specify which format it is using. application/x-www-form-urlencoded and application/json are common values.

Chapter 7, Building a Simple Web Server

1. **How does an HTTP client indicate that it has finished sending the HTTP request?**

The HTTP request should end with a blank line.

2. **How does an HTTP client know what type of content the HTTP server is sending?**

The HTTP server should identify the content with a Content-Type header.

3. **How can an HTTP server identify a file's media type?**

A common method of identifying a file's media type is just to look at the file extension. The server is free to use other methods though. When sending dynamic pages or data from a database, there will be no file and therefore no file extension. In this case, the server must know the media type from its context.

4. **How can you tell whether a file exists on the filesystem and is readable by your program? Is** fopen(filename, "r") != 0 **a good test?**

This is not a trivial problem. A robust program will need to consider system specific APIs carefully. Windows uses special filenames that will trip up a program that relies only on fopen() to check for a file's existence.

Chapter 8, Making Your Program Send Email

1. **What port does SMTP operate on?**

   SMTP does mail transmission over TCP port 25. Many providers use alternative ports for mail submission.

2. **How do you determine which SMTP server receives mail for a given domain?**

   The mail servers responsible for receiving mail for a given domain are given by MX-type DNS records.

3. **How do you determine which SMTP server sends mail for a given provider?**

   It's not possible to determine that in the general case. In any case, several servers could be responsible. Sometimes these servers will be listed under a TXT-type DNS record using SPF, but that is certainly not universal.

4. **Why won't an SMTP server relay mail without authentication?**

   Open relay SMTP servers are targeted by spammers. SMTP servers require authentication to prevent abuse.

5. **How are binary files sent as email attachments when SMTP is a text-based protocol?**

   Binary files must be re-encoded as plain text. The most common method is with `Content-Transfer-Encoding: base64`.

Chapter 9, Loading Secure Web Pages with HTTPS and OpenSSL

1. **What port does HTTPS typically operate on?**

   HTTPS connects over TCP port 443.

2. **How many keys does symmetric encryption use?**

   Symmetric encryption uses one key. Data is encrypted and decrypted with the same key.

3. **How many keys does asymmetric encryption use?**

Asymmetric encryption use two different, but mathematically related, keys. Data is encrypted with one and decrypted with the other.

4. **Does TLS use symmetric or asymmetric encryption?**

TLS use both symmetric and asymmetric encryption algorithms to function.

5. **What is the difference between SSL and TLS?**

TLS is the successor to SSL. SSL is now deprecated.

6. **What purpose do certificates fulfill?**

Certificates allow a server or client to verify their identity.

Chapter 10, Implementing a Secure Web Server

1. **How does a client decide whether it should trust a server's certificate?**

There are various ways a client can trust a server's certificate. The chain-of-trust model is the most common. In this model, the client explicitly trusts an authority. The client then implicitly trusts any certificates it encounters that are signed by this trusted authority.

2. **What is the main issue with self-signed certificates?**

Self-signed certificates aren't signed by a trusted certificate authority. Web browsers won't know to trust self-signed certificates unless the user adds a special exception.

3. **What can cause** SSL_accept() **to fail?**

SSL_accept() fails if the client doesn't trust the server's certificate or if the client and server can't agree on a mutually supported protocol version and cipher suite.

4. **Can** `select()` **be used to multiplex connections for HTTPS servers?**

Yes, but be aware that `select()` works on the underlying TCP connection layer, not on the TLS layer. Therefore, when `select()` indicates that a socket has data waiting, it does not necessarily mean that there is new TLS data ready.

Chapter 11, Establishing SSH Connections with libssh

1. **What is a significant downside of using Telnet?**

Essentially, Telnet provides no security features. Passwords are sent as plaintext.

2. **Which port does SSH typically run on?**

SSH's official port is TCP port `22`. In practice, it is common to run SSH on arbitrary ports in an attempt to hide from attackers. With a properly secured server, these attackers are a nuisance rather than a legitimate threat.

3. **Why is it essential that the client authenticates the SSH server?**

If the client doesn't verify the SSH server's identity, then it could be tricked into sending credentials to an impostor.

4. **How is the server typically authenticated?**

SSH servers typically use certificates to identity themselves. This is similar to how servers are authenticated when using HTTPS.

5. **How is the SSH client typically authenticated?**

It is still common for clients to authenticate with a password. The downside to this method is that if a client is somehow tricked into connecting to an impostor server, then their password will be compromised. SSH provides alternate methods, including authenticating clients using certificates, that aren't susceptible to replay attacks.

Chapter 12, Network Monitoring and Security

1. **Which tool would you use to test the reachability of a target system?**

   The `ping` tool is useful to test reachability.

2. **Which tool lists the routers to a destination system?**

   The `traceroute` (`tracert` on Windows) tool will show the network path to a target system.

3. **What are raw sockets used for?**

   Raw sockets allow the programmer to specify directly what goes into a network packet. They provide lower-level access than TCP and UDP sockets, and can be used to implement additional protocols, such as ICMP.

4. **Which tools list the open TCP sockets on your system?**

   The netstat tool can be used to show open connections on your local system.

5. **What is one of the biggest concerns with security for networked C programs?**

   When programming networked applications in C, special care must be given to memory safety. Even a small mistake could allow an attacker to compromise your program.

Chapter 13, Socket Programming Tips and Pitfalls

1. **Is it ever acceptable just to terminate a program if a network error is detected?**

   Yes. For some applications terminating on error is the right call. For more substantial applications, the ability to retry and continue on may be needed.

2. **Which system functions are used to convert error codes into text descriptions?**

   You can use `FormatMessage()` on Windows and `strerror()` on other platforms to obtain error messages.

3. **How long does it take for a call to** connect() **to complete on a TCP socket?**

A call to connect() typically blocks for at least one network time round trip while the TCP three-way handshake is being completed.

4. **What happens if you call** send() **on a disconnected TCP socket?**

On Unix-based systems, your program can receive a SIGPIPE signal. It is important to plan for that. Otherwise, send() returns –1.

5. **How can you ensure that the next call to** send() **won't block?**

Either use select() to make sure the socket is ready for more data or use non-blocking sockets.

6. **What happens if both peers to a TCP connection try to send a large amount of data simultaneously?**

If both sides to a TCP connection are calling send(), but not recv(), then they can be trapped in a deadlocked state. It is important to intersperse calls to send() with calls to recv(). The use of select() can help inform your program about what to do next.

7. **Can you improve application performance by disabling the Nagle algorithm?**

It depends on what your application is doing. For real-time applications using TCP, disabling the Nagle algorithm is often a good trade-off for decreasing latency at the expense of bandwidth efficiency. For other applications, disabling it can decrease throughput, increase network congestion, and even increase latency.

8. **How many connections can** select() **handle?**

It depends on your platform. It is defined in the FD_SETSIZE macro, which is easily increased on Windows but not on other platforms. Typically, the upper limit is around 1,024 sockets.

Chapter 14, Web Programming for the Internet of Things

1. **What are the drawbacks to using Wi-Fi connectivity?**

   Wi-Fi can be difficult for end user setup. It's also not available everywhere.

2. **What are the drawbacks to using Ethernet connectivity?**

   Many devices aren't used in areas where wiring has been run.

3. **What are the drawbacks to using cellular connectivity?**

   Cellular connectivity is expensive. It can also have increased latency and larger power requirements when compared to other methods.

4. **What are some advantages to using a single-board computer with embedded Linux? What are the drawbacks?**

   Having access to a full operating system, such as Linux, can simplify software development. However, **Single-Board Computers** (**SBCs**) are relatively expensive and offer few board-level connectivity options and peripherals when compared to microcontrollers. They also require lots of power, relatively speaking.

5. **What are some advantages to using a microcontroller in your IoT device?**

   Many IoT devices will need to use a microcontroller to provide their basic functionality anyway. Microcontrollers are cheap, offer a wide range of peripherals, are able to meet real-time performance constraints, and can run on very little power.

6. **Is the use of HTTPS always appropriate in IoT device communication?**

   HTTPS is a decent way to secure IoT communication for most applications; however, it has a lot of processing and bandwidth overhead. Each application is unique, and the security scheme used should be chosen based on your exact needs.

Setting Up Your C Compiler on Windows

Microsoft Windows is one of the most popular desktop operating systems.

Before beginning, I highly recommend that you install **7-Zip** from `https://www.7-zip.org/`. 7-Zip will allow you to extract the various compression archive formats that library source code is distributed in.

Let's continue and get MinGW, OpenSSL, and `libssh` set up on Windows 10.

Installing MinGW GCC

MinGW is a port of GCC to Windows. It is the compiler we recommend for this book.

You can obtain MinGW from `http://www.mingw.org/`. Find the download link on that page and download and run the **MinGW Installation Manager** (**mingw-get**).

The MinGW Installation Manager is a GUI tool for installing MinGW. It's shown in the following screenshot:

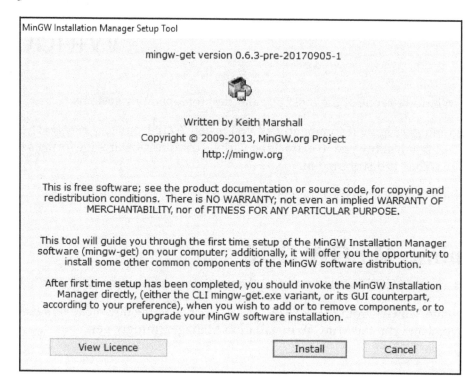

Click **Install**. Then, click **Continue**. Wait while some files download, and then click **Continue** once again.

At this point, the tool will give you a list of packages that you can install. You need to mark **mingw32-base-bin**, **msys-base-bin**, and **mingw32-gcc-g++-bin** for installation. This is shown in the following screenshot:

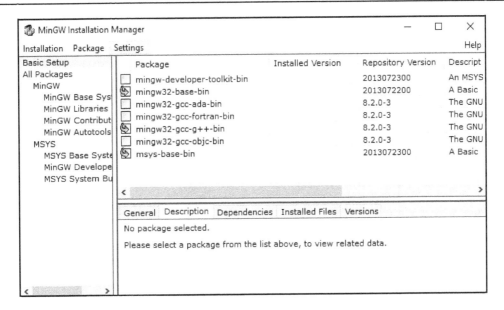

You will also want to select the **mingw32-libz-dev** package. It is listed under the **MinGW Libraries** section. The following screenshot shows this selection:

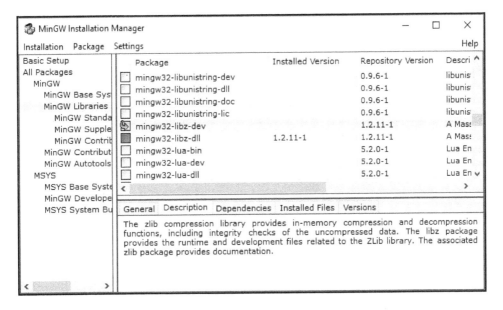

The `g++` and `libz` packages we've selected are required for building `libssh` later.

When you're ready to proceed, click **Installation** from the menu and select **Apply Changes**.

A new dialog will show the changes to be made. The following screenshot shows what this dialog may look like:

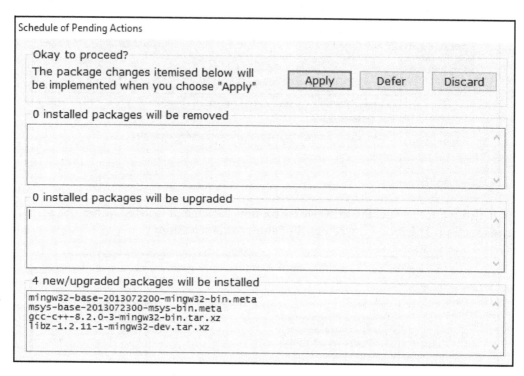

Click the **Apply** button to download and install the packages. Once the installation is complete, you can close the **MinGW Installation Manager**.

To be able to use MinGW from the command line easily, you will need to add MinGW to your PATH.

The steps for adding MinGW to your PATH are as follows:

1. Open the **System** control panel (Windows key + *Pause/Break*).
2. Select **Advanced system settings**:

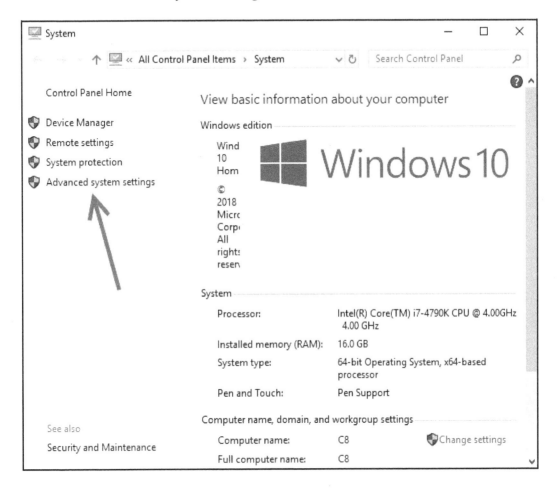

3. From the **System Properties** window, navigate to the **Advanced** tab and click the **Environment Variables...** button:

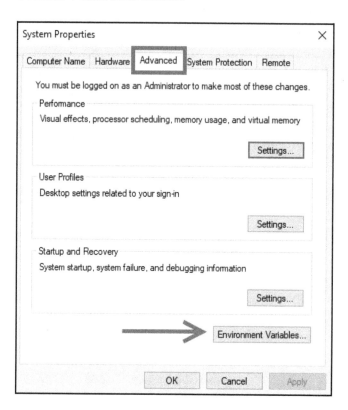

4. From this screen, find the PATH variable under **System variables**. Select it and press **Edit...**.

5. Click **New** and type in the MinGW path—C:\mingw\bin, as shown in the following screenshot:

6. Click **OK** to save your changes.

Once MinGW has been added to your PATH, you can open a new command window and enter gcc --version to ensure that gcc has been installed correctly. This is shown in the following screenshot:

```
C:\Windows\system32\cmd.exe                                    —    □    ×

Microsoft Windows [Version 10.0.10586]
(c) 2015 Microsoft Corporation. All rights reserved.

C:\Users\Tombob>gcc --version
gcc (MinGW.org GCC-8.2.0-3) 8.2.0
Copyright (C) 2018 Free Software Foundation, Inc.
This is free software; see the source for copying conditions.  There is NO
warranty; not even for MERCHANTABILITY or FITNESS FOR A PARTICULAR PURPOSE.
```

Installing Git

You will need to have the `git` version control software installed to download this book's code.

`git` is available from `https://git-scm.com/download`. A handy GUI-based installer is provided, and you shouldn't have any issues getting it working. When installing, be sure to check the option for adding `git` to your `PATH`. This is shown in the following screenshot:

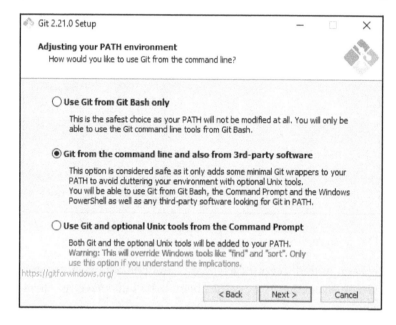

After `git` has finished installing, you can test it by opening a new command window and entering `git --version`:

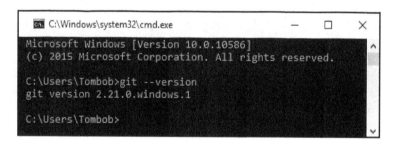

Installing OpenSSL

The OpenSSL library can be tricky to get going on Windows.

If you are brave, you can obtain the OpenSSL library source code directly from `https://www.openssl.org/source/`. You will, of course, need to build OpenSSL before it can be used. Building OpenSSL is not easy, but instructions are provided in the `INSTALL` and `NOTES.WIN` files included with the OpenSSL source code.

An easier alternative is to install prebuilt OpenSSL binaries. You can find a list of prebuilt OpenSSL binaries from the OpenSSL wiki at `https://wiki.openssl.org/index.php/Binaries`. You will need to locate binaries that match your operating system and compiler. Installing them will be a matter of copying the relevant files to the MinGW `include`, `lib`, and `bin` directories.

The following screenshot shows a binary OpenSSL distribution. The `include` and `lib` folders should be copied over to `c:\mingw\` and merged with the existing folders, while `openssl.exe` and the two DLL files need to be placed in `c:\mingw\bin\`:

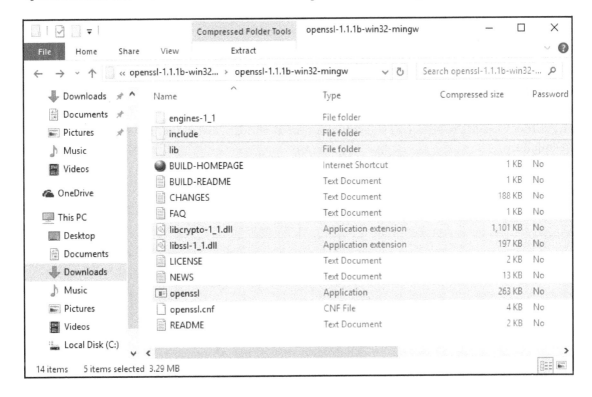

You can try building `openssl_version.c` from Chapter 9, *Loading Secure Web Pages with HTTPS and OpenSSL,* to test that everything is installed correctly. It should look like the following:

```
C:\Windows\system32\cmd.exe                                        —    □    ×

C:\Hands-on-network-programming-with-c\chap09>gcc openssl_version.c -o openssl_version.exe -lcrypto

C:\Hands-on-network-programming-with-c\chap09>openssl_version.exe
OpenSSL version: OpenSSL 1.1.1b  26 Feb 2019

C:\Hands-on-network-programming-with-c\chap09>
```

Installing libssh

You can obtain the latest `libssh` library from `https://www.libssh.org/`. If you are proficient in installing C libraries, feel free to give it a go. Otherwise, read on for step-by-step instructions.

Before beginning, be sure that you've first installed the OpenSSL libraries successfully. These are required by the `libssh` library.

We will need CMake installed in order to build `libssh`. You can obtain CMake from `https://cmake.org/`. They provide a nice GUI installer, and you shouldn't run into any difficulties. Make sure you select the option to add CMake to your `PATH` during installation:

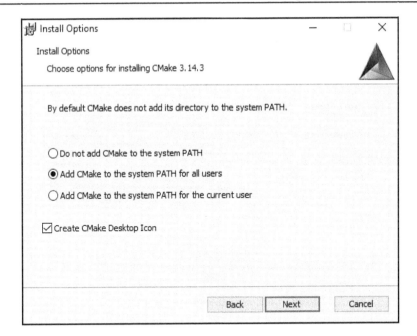

Once you have the CMake tool and the OpenSSL libraries installed, navigate to the `libssh` website to download the `libssh` source code. At the time of writing, Version 0.8.7 is the latest, and it is available from `https://www.libssh.org/files/0.8/`. Download and extract the `libssh` source code.

Take a look at the included `INSTALL` file.

Now, open a command window in the `libssh` source code directory. Create a new `build` folder with the following commands:

```
mkdir build
cd build
```

Keep this command window open. We'll do the build here in a minute.

Start **CMake 3.14.3 (cmake-gui)** from the start menu or desktop shortcut.

You need to set the source code and build locations using the **Browse Source...** and **Browse Build...** buttons. This is shown in the following screenshot:

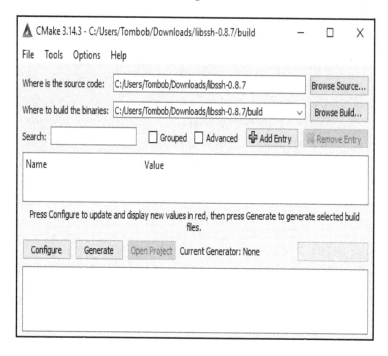

Then, click **Configure**.

On the next screen, select **MinGW Makefiles** as the generator for this project. Click **Finish**.

It may take a moment to process.

From the configuration options, make the following changes:

1. Uncheck **WITH_NACL**
2. Uncheck **WITH_GSSAPI**
3. Change `CMAKE_INSTALL_PREFIX` to `c:\mingw`

Then, click **Configure** again. It will take a moment. If everything worked, click **Generate**.

You should now be able to build `libssh`.

Go back to your command window in the build directory. Use the following command to complete the build:

```
mingw32-make
```

After the build completes, use the following command to copy the files over to your MinGW installation:

```
mingw32-make install
```

You can try building `ssh_version.c` from Chapter 11, *Establishing SSH Connections with libssh*, to test that everything is installed correctly. It should look like the following:

Alternatives

In this book, we recommend free software whenever possible. This is important for user freedom, and this is one reason we recommend GCC throughout the book.

In addition, to MinGW GCC, the Clang C compiler is also open source and excellent quality. The code in this book was also tested to run successfully using Clang on Windows.

Command-line tools such as GCC and Clang are often easier to integrate into the complicated workflows required for larger projects. These open source tools also provide better standards compliance than Microsoft's compilers.

That said, the code in this book also works with Microsoft's compilers. The code was tested for both Microsoft Visual Studio 2015 and Microsoft Visual Studio 2017.

Setting Up Your C Compiler on Linux

Linux is an excellent choice for C programming. It has arguably the easiest setup and the best support for C programming out of the three operating systems covered in this book.

Using Linux also allows you to take the ethical high road and feel good about supporting free software.

One issue with describing the setup for Linux is that there are many Linux distributions with different software. In this appendix, we will provide the commands needed to set up on systems using the `apt` package manager, such as **Debian Linux** and **Ubuntu Linux**. If you are using a different Linux distribution, you will need to find the commands relevant to your system. Refer to your distribution's documentation for help.

Before diving right in, take a moment to make sure your package list is up to date. This is done with the following command:

```
sudo apt-get update
```

With `apt` ready to go, setup is easy. Let's get started.

Installing GCC

The first step is to get the C compiler, `gcc`, installed.

Assuming your system uses `apt` as its package manager, try the following commands to install `gcc` and prepare your system for C programming:

```
sudo apt-get install build-essential
```

Once the `install` command completes, you should be able to run the following command to find the version of `gcc` that was installed:

```
gcc --version
```

Installing Git

You will need to have the Git version control software installed to download this book's code.

Assuming your system uses the `apt` package manager, you can install Git with the following command:

```
sudo apt-get install git
```

Check whether Git has installed successfully by means of the following command:

```
git --version
```

Installing OpenSSL

OpenSSL can be tricky. You can try your distribution's package manager with the following commands:

```
sudo apt-get install openssl libssl-dev
```

The problem is that your distribution may have an old version of OpenSSL. If that is the case, you should obtain the OpenSSL libraries directly from `https://www.openssl.org/ source/`. You will, of course, need to build OpenSSL before it can be used. Building OpenSSL is not easy, but instructions are provided in the `INSTALL` file included with the OpenSSL source code. Note that its build system requires that you have **Perl** installed.

Installing libssh

You can try installing `libssh` from your package manager with the following command:

```
sudo apt-get install libssh-dev
```

The problem is that the code in this book is not compatible with older versions of `libssh`. Therefore, I recommend you build `libssh` yourself.

You can obtain the latest `libssh` library from `https://www.libssh.org/`. If you are proficient in installing C libraries, feel free to give it a go. Otherwise, read on for the step-by-step instructions.

Before beginning, be sure that you've first installed the OpenSSL libraries successfully. These are required by the `libssh` library.

We will also need CMake installed in order to build `libssh`. You can obtain CMake from `https://cmake.org/`. You can also get it from your distro's packaging tool with the following command:

```
sudo apt-get install cmake
```

Finally, the `zlib` library is also required by `libssh`. You can install the `zlib` library using this command:

```
sudo apt-get install zlib1g-dev
```

Once you have CMake, the `zlib` library, and the OpenSSL library installed, locate the version of `libssh` you would like from `https://www.libssh.org/`. Version 0.8.7 is the latest at the time of writing. You can download and extract the `libssh` source code with the following commands:

```
wget https://www.libssh.org/files/0.8/libssh-0.8.7.tar.xz
tar xvf libssh-0.8.7.tar.xz
cd libssh-0.8.7
```

I recommend that you take a look at the installation instructions included with `libssh`. You can use `less` to view them. Press the *Q* key to quit `less`:

```
less INSTALL
```

Once you've familiarized yourself with the build instructions, you can try building `libssh` with these commands:

```
mkdir build
cd build
cmake ..
make
```

The final step is to install the library with the following command:

```
sudo make install
```

Setting Up Your C Compiler on macOS

macOS can be a good development environment for a C programmer. Let's get started.

Installing Homebrew and the C compiler

Setup on macOS will be simplified greatly if we use the **Homebrew** package manager.

The Homebrew package manager makes installing a C compiler and development libraries much easier than it would otherwise be.

To install Homebrew, navigate your web browser to `https://brew.sh/`:

The website gives the command to install Homebrew. You will need to open a new Terminal window and paste in the command:

```
● ● ●                    ⌂ honp — -bash — 80×29
[honps-Mac:~ honp$ /usr/bin/ruby -e "$(curl -fsSL https://raw.githubusercontent.c▤
om/Homebrew/install/master/install)"
==> This script will install:
/usr/local/bin/brew
/usr/local/share/doc/homebrew
/usr/local/share/man/man1/brew.1
/usr/local/share/zsh/site-functions/_brew
/usr/local/etc/bash_completion.d/brew
/usr/local/Homebrew
==> The following new directories will be created:
/usr/local/bin
/usr/local/etc
/usr/local/include
/usr/local/lib
/usr/local/sbin
/usr/local/share
/usr/local/var
/usr/local/opt
/usr/local/share/zsh
/usr/local/share/zsh/site-functions
/usr/local/var/homebrew
/usr/local/var/homebrew/linked
/usr/local/Cellar
/usr/local/Caskroom
/usr/local/Homebrew
/usr/local/Frameworks
==> The Xcode Command Line Tools will be installed.

Press RETURN to continue or any other key to abort
```

Just follow the instructions until Homebrew has finished installing.

Installing Homebrew will also cause the Xcode command-line tools to install. This means you'll have a C compiler ready to go. Test that you have a working C compiler with the gcc --version command:

```
● ◐ ◯                    ⌂ honp — -bash — 80×9
honps-Mac:~ honp$ gcc --version
Configured with: --prefix=/Library/Developer/CommandLineTools/usr --with-gxx-inc
lude-dir=/Library/Developer/CommandLineTools/SDKs/MacOSX10.14.sdk/usr/include/c+
+/4.2.1
Apple LLVM version 10.0.1 (clang-1001.0.46.4)
Target: x86_64-apple-darwin18.2.0
Thread model: posix
InstalledDir: /Library/Developer/CommandLineTools/usr/bin
honps-Mac:~ honp$ ▊
```

Note that macOS installs the Clang compiler, but aliases it as GCC. In any case, it will work fine for our purposes.

Git should also have been installed. You can verify this with `git --version`:

```
● ◐ ◯                    ⌂ honp — -bash — 80×9
honps-Mac:~ honp$ git --version
git version 2.20.1 (Apple Git-117)
honps-Mac:~ honp$ ▊
```

Installing OpenSSL

Assuming you already have Homebrew installed, installing OpenSSL is easy.

Open a new Terminal window and use the following command to install the OpenSSL library:

```
brew install openssl@1.1
```

At the time of writing, the default Homebrew `openssl` package is outdated, therefore we'll use the `openssl@1.1` package instead.

The following screenshot shows Homebrew installing the `openssl@1.1` package:

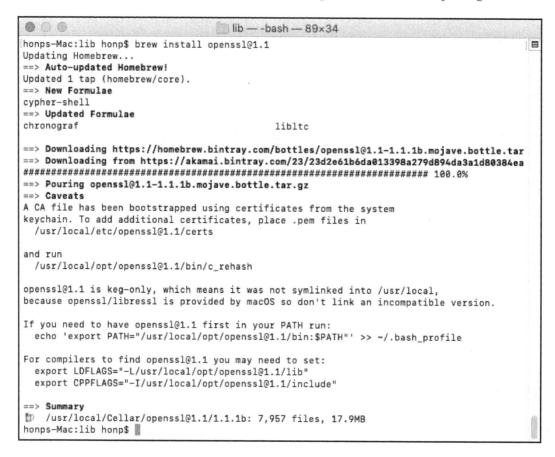

```
● ○ ○                          📁 lib — -bash — 89×34
[honps-Mac:lib honp$ brew install openssl@1.1
Updating Homebrew...
==> Auto-updated Homebrew!
Updated 1 tap (homebrew/core).
==> New Formulae
cypher-shell
==> Updated Formulae
chronograf                                    libltc

==> Downloading https://homebrew.bintray.com/bottles/openssl@1.1-1.1.1b.mojave.bottle.tar
==> Downloading from https://akamai.bintray.com/23/23d2e61b6da013398a279d894da3a1d80384ea
######################################################################## 100.0%
==> Pouring openssl@1.1-1.1.1b.mojave.bottle.tar.gz
==> Caveats
A CA file has been bootstrapped using certificates from the system
keychain. To add additional certificates, place .pem files in
  /usr/local/etc/openssl@1.1/certs

and run
  /usr/local/opt/openssl@1.1/bin/c_rehash

openssl@1.1 is keg-only, which means it was not symlinked into /usr/local,
because openssl/libressl is provided by macOS so don't link an incompatible version.

If you need to have openssl@1.1 first in your PATH run:
  echo 'export PATH="/usr/local/opt/openssl@1.1/bin:$PATH"' >> ~/.bash_profile

For compilers to find openssl@1.1 you may need to set:
  export LDFLAGS="-L/usr/local/opt/openssl@1.1/lib"
  export CPPFLAGS="-I/usr/local/opt/openssl@1.1/include"

==> Summary
🍺  /usr/local/Cellar/openssl@1.1/1.1.1b: 7,957 files, 17.9MB
honps-Mac:lib honp$ █
```

Be sure to read any output from `brew`.

The method suggested by the `brew` command has you pass `-L` and `-I` to the compiler to tell it where to locate the OpenSSL libraries. This is tedious.

You may prefer to symlink the installed library files to the `/usr/local` path so that your compiler can find them automatically. Whether you want to do this or not is up to you.

If you want to create symlinks so the compiler can find OpenSSL, try the following commands:

```
cd /usr/local/include
ln -s ../opt/openssl@1.1/include/openssl .
cd /usr/local/lib
for i in ../opt/openssl@1.1/lib/lib*; do ln -vs $i .; done
```

This is shown in the following screenshot:

```
                          lib — -bash — 89×12
honps-Mac:lib honp$ cd /usr/local/include
honps-Mac:include honp$ ln -s ../opt/openssl@1.1/include/openssl .
honps-Mac:include honp$ cd /usr/local/lib
honps-Mac:lib honp$ for i in ../opt/openssl@1.1/lib/lib*; do ln -vs $i .; done
./libcrypto.1.1.dylib -> ../opt/openssl@1.1/lib/libcrypto.1.1.dylib
./libcrypto.a -> ../opt/openssl@1.1/lib/libcrypto.a
./libcrypto.dylib -> ../opt/openssl@1.1/lib/libcrypto.dylib
./libssl.1.1.dylib -> ../opt/openssl@1.1/lib/libssl.1.1.dylib
./libssl.a -> ../opt/openssl@1.1/lib/libssl.a
./libssl.dylib -> ../opt/openssl@1.1/lib/libssl.dylib
honps-Mac:lib honp$
```

You can try building `openssl_version.c` from Chapter 9, *Loading Secure Web Pages with HTTPS and OpenSSL*, to test that everything is installed correctly. It should look like the following:

```
                     chap09 — -bash — 80×5
honps-Mac:chap09 honp$ gcc openssl_version.c -o openssl_version -lcrypto
honps-Mac:chap09 honp$ ./openssl_version
OpenSSL version: OpenSSL 1.1.1b  26 Feb 2019
honps-Mac:chap09 honp$
```

Installing libssh

With Homebrew installed, installing `libssh` is also very easy.

Open a new Terminal window and use the following command to install the OpenSSL library:

```
brew install libssh
```

This is shown in the following screenshot:

```
● ◐ ●                    🏠 honp — -bash — 89×9
[honps-Mac:~ honp$ brew install libssh
Updating Homebrew...
==> Downloading https://homebrew.bintray.com/bottles/libssh-0.8.7.mojave.bottle.tar.gz
==> Downloading from https://akamai.bintray.com/68/68027b65dae117340cbaa3b72f09850322b16a
################################################################### 100.0%
==> Pouring libssh-0.8.7.mojave.bottle.tar.gz
🍺  /usr/local/Cellar/libssh/0.8.7: 21 files, 1.3MB
honps-Mac:~ honp$
```

This time, we don't need to mess with any other options. You can start using libssh right away.

You can try building ssh_version.c from Chapter 11, *Establishing SSH Connections with libssh*, to test that everything is installed correctly. It should look like the following:

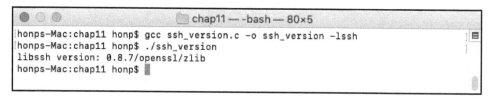

```
● ◐ ●                    chap11 — -bash — 80×5
[honps-Mac:chap11 honp$ gcc ssh_version.c -o ssh_version -lssh
[honps-Mac:chap11 honp$ ./ssh_version
libssh version: 0.8.7/openssl/zlib
honps-Mac:chap11 honp$
```

That concludes the setup for macOS.

Example Programs

This book's code repository, located at `https://github.com/codeplea/hands-on-network-programming-with-c,` includes 44 example programs. These programs are explained in detail throughout the book.

Code license

The example programs provided in this book's code repository are released under the MIT license, the text of which follows:

```
Permission is hereby granted, free of charge, to any person obtaining
a copy of this software and associated documentation files (the
"Software"), to deal in the Software without restriction, including without
limitation the rights to use, copy, modify, merge, publish, distribute,
sublicense, and/or sell copies of the Software, and to permit persons to
whom the Software is furnished to do so, subject to the following
conditions:

The above copyright notice and this permission notice shall be included
in all copies or substantial portions of the Software.

THE SOFTWARE IS PROVIDED "AS IS", WITHOUT WARRANTY OF ANY KIND, EXPRESS
OR IMPLIED, INCLUDING BUT NOT LIMITED TO THE WARRANTIES OF MERCHANTABILITY,
FITNESS FOR A PARTICULAR PURPOSE AND NONINFRINGEMENT. IN NO EVENT SHALL THE
AUTHORS OR COPYRIGHT HOLDERS BE LIABLE FOR ANY CLAIM, DAMAGES OR OTHER
LIABILITY, WHETHER IN AN ACTION OF CONTRACT, TORT OR OTHERWISE, ARISING
FROM, OUT OF OR IN CONNECTION WITH THE SOFTWARE OR THE USE OR OTHER
DEALINGS IN THE SOFTWARE.
```

Code included with this book

The following is a list of the 44 example programs included with this book, by chapter.

Chapter 1 – Introducing Networks and Protocols

This chapter includes the following example programs:

- win_init.c: Example code to initialize Winsock (Windows only)
- win_list.c: Lists all local IP addresses (Windows only)
- unix_list.c: Lists all local IP addresses (Linux and macOS only)

Chapter 2 – Getting to Grips with Socket APIs

This chapter includes the following example programs:

- sock_init.c: Example program to include all necessary headers and initialize
- time_console.c: Prints the current date and time to the console
- time_server.c: Serves a web page giving the current date and time
- time_server_ipv6.c: The same as the preceding code but listens for IPv6 connections
- time_server_dual.c: The same as the preceding code but listens for IPv6/IPv4 dual-stack connections

Chapter 3 – An In-Depth Overview of TCP Connections

This chapter includes the following example programs:

- tcp_client.c: Establishes a TCP connection and sends/receives data from the console.
- tcp_serve_toupper.c: A TCP server servicing multiple connections using select(). Echoes received data back to the client all in uppercase.

- `tcp_serve_toupper_fork.c`: The same as the preceding code but uses `fork()` instead of `select()`. (Linux and macOS only.)
- `tcp_serve_chat.c`: A TCP server that relays received data to every other connected client.

Chapter 4 – Establishing UDP Connections

This chapter includes the following example programs:

- `udp_client.c`: Sends/receives UDP data from the console
- `udp_recvfrom.c`: Uses `recvfrom()` to receive one UDP datagram
- `udp_sendto.c`: Uses `sendto()` to send one UDP datagram
- `udp_serve_toupper.c`: Listens for UDP data and echoes it back to the sender all in uppercase
- `udp_serve_toupper_simple.c`: The same as the preceding code but doesn't use `select()`

Chapter 5 – Hostname Resolution and DNS

This chapter includes the following example programs:

- `lookup.c`: Uses `getaddrinfo()` to look up addresses for a given hostname
- `dns_query.c`: Encodes and sends a UDP DNS query, then listens for and decodes the response

Chapter 6 – Building a Simple Web Client

This chapter includes the following example program:

- `web_get.c`: A minimal HTTP client to download a web resource from a given URL

Chapter 7 – Building a Simple Web Server

This chapter includes the following example programs:

- `web_server.c`: A minimal web server capable of serving a static website
- `web_server2.c`: A minimal web server (no globals)

Chapter 8 – Making Your Program Send Email

This chapter includes the following example program:

- `smtp_send.c`: A simple program to transmit an email

Chapter 9 – Loading Secure Web Pages with HTTPS and OpenSSL

The examples in this chapter use OpenSSL. Be sure to link to the OpenSSL libraries when compiling (`-lssl -lcrypto`):

- `openssl_version.c`: A program to report the installed OpenSSL version
- `https_simple.c`: A minimal program that requests a web page using HTTPS
- `https_get.c`: The HTTP client of `Chapter 6`, *Building a Simple Web Client*, modified to use HTTPS
- `tls_client.c`: The TCP client program of `Chapter 3`, *An In-Depth Overview of TCP Connections*, modified to use TLS
- `tls_get_cert.c`: Prints a certificate from a TLS server

Chapter 10 – Implementing a Secure Web Server

The examples in this chapter use OpenSSL. Be sure to link to the OpenSSL libraries when compiling (`-lssl -lcrypto`):

- `tls_time_server.c`: The time server of `Chapter 2`, *Getting to Grips with Socket APIs*, modified to use HTTPS
- `https_server.c`: The web server of `Chapter 7`, *Building a Simple Web Server*, modified to use HTTPS

Chapter 11 – Establishing SSH Connections with libssh

The examples in this chapter use libssh. Be sure to link to the libssh libraries when compiling (-lssh):

- ssh_version.c: A program to report the libssh version
- ssh_connect.c: A minimal client that establishes an SSH connection
- ssh_auth.c: A client that attempts SSH client authentication using a password
- ssh_command.c: A client that executes a single remote command over SSH
- ssh_download.c: A client that downloads a file over SSH/SCP

Chapter 12 – Network Monitoring and Security

This chapter doesn't include any example programs.

Chapter 13 – Socket Programming Tips and Pitfalls

This chapter includes the following example programs:

- connect_timeout.c: Shows how to time out a connect() call early.
- connect_blocking.c: For comparison with connect_timeout.c.
- server_reuse.c: Demonstrates the use of SO_REUSEADDR.
- server_noreuse.c: For comparison with server_reuse.c.
- server_crash.c: This server purposefully writes to a TCP socket after the client disconnects.
- error_text.c: Shows how to obtain error code descriptions.
- big_send.c: A TCP client that sends lots of data after connecting. Used to show the blocking behavior of send().
- server_ignore.c: A TCP server that accepts connections, then simply ignores them. Used to show the blocking behavior of send().
- setsize.c: Shows the maximum number of sockets select() can handle.

Chapter 14 – Web Programming for the Internet of Things

This chapter doesn't include any example programs.

Other Book You May Enjoy

If you enjoyed this book, you may be interested in another book by Packt:

C Programming Cookbook
B.M. Harwani

ISBN: 9781789617450

- Manipulate single and multi-dimensional arrays
- Perform complex operations on strings
- Understand how to use pointers and memory optimally
- Discover how to use arrays, functions, and strings to make large applications
- Implement multitasking using threads and process synchronization
- Establish communication between two or more processes using different techniques
- Store simple text in files and store data in a database

Leave a review - let other readers know what you think

Please share your thoughts on this book with others by leaving a review on the site that you bought it from. If you purchased the book from Amazon, please leave us an honest review on this book's Amazon page. This is vital so that other potential readers can see and use your unbiased opinion to make purchasing decisions, we can understand what our customers think about our products, and our authors can see your feedback on the title that they have worked with Packt to create. It will only take a few minutes of your time, but is valuable to other potential customers, our authors, and Packt. Thank you!

Index